Global Warming and the Wine Industry - Challenges, Innovations and Future Prospects

Edited by Fernanda Cosme,
Fernando M. Nunes and Luís Filipe-Ribeiro

Published in London, United Kingdom

Global Warming and the Wine Industry - Challenges, Innovations and Future Prospects
http://dx.doi.org/10.5772/intechopen.111161
Edited by Fernanda Cosme, Fernando M. Nunes and Luís Filipe-Ribeiro

Contributors
Camelia Elena Luchian, Christian Poulsen, Elena Cristina Scutarașu, Fernanda Cosme, Fernando M. Nunes, Ionica Bosoi, Komal Sekhon, Liliana Rotaru, Lucia Cintia Colibaba, Luís Filipe-Ribeiro, Mette Mønsted, Miguel Ribeiro, Mihai Cristian Focea, Mărioara Pușcalău, Qun Sun, Valeriu Cotea, Valeriu V. Cotea

First published in London, United Kingdom, 2024 by IntechOpen
IntechOpen is the global imprint of INTECHOPEN LIMITED, registered in England and Wales, registration number: 11086078, 167-169 Great Portland Street, London, W1W 5PF, United Kingdom

British Library Cataloguing-in-Publication Data
A catalogue record for this book is available from the British Library

Additional hard and PDF copies can be obtained from orders@intechopen.com

Global Warming and the Wine Industry - Challenges, Innovations and Future Prospects
Edited by Fernanda Cosme, Fernando M. Nunes and Luís Filipe-Ribeiro
p. cm.
Print ISBN 978-0-85466-272-2
Online ISBN 978-0-85466-271-5
eBook (PDF) ISBN 978-0-85466-273-9

For EU product safety concerns:
IN TECH d.o.o., Prolaz Marije Krucifikse Kozulić 3, 51000 Rijeka, Croatia,
info@intechopen.com or visit our website at intechopen.com.

Meet the editors

Fernanda Cosme is an associate professor at the University of Trás-os-Montes and Alto Douro. She has as a main research lines grapes and wines in respect to grape composition, wine technology, wine stability, quality and safety, and aging. Recently, she edited three books on enology as a scientific editor. She is the author of patents and articles in international journals cited in the Science Citation Index. She has also published book chapters and articles in open-access journals with peer review and participates regularly in international and national congresses as a speaker. She has participated in several funded research projects, including projects with industrial partners, and supervised Ph.D. and MSc student theses in food science and enology (orcid.org/0000-0001-5924-1050; Scopus ID: 24437875500).

Fernando M. Nunes is an associate professor with habilitation at the University of Trás-os-Montes and Alto Douro in the area of food chemistry and analytical chemistry. He has experience in the analysis of phenolic compounds and aromas in wine, carbohydrate analysis, Maillard reactions, and melanoidins. He is the author of 18 patents and 177 articles in international journals cited in the Science Citation Index in the areas of food chemistry, carbohydrate chemistry, polysaccharide chemistry, phenolic compounds, and biological activities of these compounds. He has been the principal investigator of several research projects with industrial partners (orcid.org/0000-0001-5540-318X; Scopus ID 7102392847).

Luís Filipe-Ribeiro has a master's degree in viticulture and enology, a doctorate in chemical and biological sciences, with a postdoctorate degree in oenological sciences. He has been an enology professor since 1997–1998 and participated in many subjects (vinification, wine stabilization, analysis and analytical control of wines, sensorial analysis, etc.). He is a researcher in the Chemistry Research Centre of Vila Real (UTAD) in the Food and Wine Chemistry Laboratory, where he participates in research projects and student orientations. He is the technical director of SAI Enology, where he is responsible for the R&D+i team. Luís Filipe-Ribeiro is the author of several scientific (SCI) publications, patents, books, and book chapters. He participates regularly in international congresses. He was a consultant in viticulture and enology in several wineries (orcid.org/0000-0002-6744-4082).

Contents

Preface

Nowadays, climate change is a very important and high-priority subject worldwide due to its significant impact on all agricultural productions and various strategic sectors. That means that grape growers and wine producers must be attentive to these rapid changes and their consequences. The production of grapes and grape products is already feeling these fast changes and their important impacts, necessitating strategies to avoid or minimize the side effects of global warming by grape growers and the wine industry.

This book aims to provide important information about the impact of climate change on grapes and grape product mycotoxins and on some physicochemical parameters, such as sugar levels in grapes and grape products, dealcoholization treatments, the production of wines with low alcohol content, and innovation in the wine industry. Thus, the chapters of this book cover different topics related to the impact, strategies, and treatments to minimize the effect of climate change on grapes and grapes products.

The general impact of climate change is presented and discussed in the first introductory chapter of this book. The fungal activities on vines and grapes are strictly dependent on the climate and microclimate conditions during their production. Climate change could alter fungal activity and the content of mycotoxins. Chapter 2 is a review that explores the presence of mycotoxins in grapes and grape products, focusing on various types such as ochratoxin A, aflatoxins, fumonisins, patulin, and others. The discussion encompasses multifaceted factors influencing mycotoxin occurrence, including environmental aspects, agricultural practices, post-harvest handling, and advanced techniques for mycotoxin detection. Mitigation strategies, such as the implementation of good agricultural practices and good manufacturing practices, are also presented and discussed.

As shown in many studies, temperature changes influence the spread of grapevine cultivation, the timing and progression of vegetation phenophases, and the overall quality of grape production. The temperature changes recorded and their influence on grapevine phenology and wine characteristics highlight the need for ongoing research and proactive measures to ensure long-term sustainability and resilience of grape cultivation in the face of climate change. These points are thoroughly explored and discussed in Chapter 3, where the question is posed, "Grape Technology vs. Climate Change: A Success Story or a Nightmare?

Recently, research on the adverse effects of alcohol consumption has spurred a trend toward low-alcohol wine, typically containing less than 8.5% alcohol by volume. This caters to health-conscious consumers and presents an economic opportunity for winemakers in an emerging market. Despite the industry's millennia-old history, there remains ample room for innovation in low-alcohol winemaking. With shifting consumer preferences and climate change, the demand for lower-alcohol wines is poised to grow, necessitating ongoing research and innovative practices to create well-balanced wines. The opportunities and challenges for low-alcohol wines are interestingly explored in Chapter 4 of this book.

Chapter 5 explores strategies to reduce the effects of global warming on sparkling wine production. The quality and characteristics of sparkling wines, their stability, and their sensory properties largely depend on the physicochemical composition of the grapes and the respective base wine, the production technology applied, and the environmental conditions. Several techniques can be implemented to produce low-alcohol base wines, and reverse osmosis is a procedure that has been successfully used in recent times to reduce the alcohol concentration while having a low negative impact on the wine composition under certain conditions. This chapter looks at the effects of reverse osmosis and the implications of inoculated yeasts on sparkling wine quality.

Finally, in the last chapter (Chapter 6), the authors explore the experimentation and practices of innovation that can take place in the wineries at the individual level and at different stages of wine production. The innovation processes are not radical but tend to be entrepreneurial practices of experimentation. The chapter takes a first glance at how practice theory can be implemented in the study of wine and entrepreneurship. The chapter is based on qualitative interviews conducted in New South Wales (Australia), Var (France), and Ribera del Duero and Toro (Spain).

Summing up, this book covers a wide range of issues and hot research topics about the impact of global warming on grapes and grape products. The challenges faced by vineyards and wine production under new climate conditions are important hot points. This compact book provides a rapid overview of the general consequences of climate change and explores some strategies and solutions to avoid or mitigate the negative impact of warm temperatures on grapes and grape products. This book is heartily recommended for graduate and Ph.D. students in enology and food science, as well as for winemakers, viticulture specialists, winegrowers, and wine producers. This book is the result of many collaborating efforts, and we gratefully acknowledge all the authors who contributed to this book and the IntechOpen team for this opportunity.

Fernanda Cosme, Fernando M. Nunes and Luís Filipe-Ribeiro
Chemistry Research Centre—Vila Real (CQ-VR),
Food and Wine Chemistry Laboratory (FoodWin),
University of Trás-os-Montes and Alto Douro,
Vila Real, Portugal

Chapter 1

Introductory Chapter: Impact of Climate Change on Grapes and Grape Products

Fernanda Cosme, Luís Filipe-Ribeiro and Fernando M. Nunes

1. Introduction

Botanically, grapes are categorized as berries, produced by the plant of the genus *Vitis*. Grapes (*Vitis* spp.) are among the most highly prized fruits worldwide. The common grapevine, indigenous to the Mediterranean region and Central Europe, belongs to the species *Vitis vinifera*, which contains somewhere between 5.000 and 10.000 distinct varieties and is one of the most widely grown and economically important grapevine species in the world, mainly as a consequence of the high quality of its wines [1]. There are approximately 60 grape species, with the majority located in the temperate zones of the Northern Hemisphere, with nearly equal distribution between Asia and America [2]. Grapes constitute a fruit that has been cultivated and consumed in numerous ways since ancient times. The term "grape products" refers to the grape berry itself, which can be consumed either fresh or after drying, as well as the range of products that can be produced from it, such as wine, fortified wine, sparkling wine, grape juice, distillates, vinegar, dried grapes, jams, jellies, and seed oil [3].

Grapes are cultivated all over the world with the world vineyard surface area estimated to be 7.3 million hectares in 2022, with vines for all purposes (wine, juice, table grapes, and dried grapes) [4]. In 2022, the world production of grapes was 80.1 million tons, of which 50% were wine grapes, 42% table grapes, and 8% dried grapes [4]. In the same year, global wine production was estimated at 258 million hectoliters, with the three main wine producers being Italy, France, and Spain [4].

In recent decades, growing attention has been given to global climate change, its associated risks, and the necessary steps for its mitigation. Given the significance of grapes for wine production in many countries and the current circumstances of climate change in regions with a tradition of wine production, it is necessary to identify strategies to maintain the alcoholic concentration of wine at a moderate level to avoid altering the wine composition and freshness concerning consumer preferences. On the other hand, climate change could also increase the incidence of mycotoxins produced by certain fungi in wine. In response, both the wine industry and the scientific community have proposed numerous techniques or methods to mitigate the adverse impacts of global warming on grapes and wines. By those reasons, this introductory chapter aims to provide a brief overview of the impact of climate change on grape quality and the incidence of mycotoxin-producing fungi, as well as strategies to mitigate these occurrences.

IntechOpen

2. Impact of climate change on grape and grape product quality

2.1 Reducing the effects of climate changes on grape and wine quality

A primary consequence of rising temperatures due to climate changes is an acceleration in the vegetative and reproductive cycles of grapevines. Grape berries are susceptible to heat stress, affecting both berry composition and wine quality. In addition to influencing primary metabolites such as organic acids, particularly malic acid [5], sugars [6–8], and amino acids [9], climate change also affects secondary metabolites, including flavonoids and aroma precursor pathways, as evidenced by transcriptomic, proteomic, and metabolomic approaches [9–16], thereby affecting the balance of berry quality-related compounds at ripeness [7, 17], which are responsible for color, bitterness, mouthfeel, and important sensory attributes. While increased temperature consistently reduces anthocyanins levels [15, 18], moderate increases in berry temperatures, achieved using open-top chambers, can decouple the anthocyanins and sugar contents in red varieties [17] and influence berry sensory characteristics [19]. Therefore, optimal sugar levels are not always correlated with similar maturity in flavor, color, or aroma [17, 19]. This has led to the suggestion that "sugar ripeness" is no longer coordinated with "phenolic" ripeness or phenolic maturation of the grapes as in warmer years, anthocyanins and tannins do not mature to the same extent as they would in cooler years. Berry acidity, another key quality parameter, is likely to decline in response to warming [20]. Warm nights can lead to the respiratory loss of malic acid, affecting the sugar-acid and aroma-acid balance, making it suboptimal [21–23].

A consequence of climate change for the wine industry is a clear increase in the sugar content of grapes and the corresponding alcohol content in wines, observed in many wine regions [24]. In Bordeaux (France), wines typically 12.5% alcohol in the 1980s have now risen to 16%, and these wines will also lack freshness and aromatic complexity [25]. For instance, in Languedoc (France), alcohol content has risen from 11 to 14%, pH has increased from 3.50 to 3.75, and total acidity has dropped from 6.0 to 4.5 g/L [26]. Wines with elevated alcohol levels are generally considered less healthy [27, 28]. Given societal concerns related to high alcohol content and the growing trend for healthier lifestyles, consumers are increasingly interested in wines with lower alcohol levels [29]. However, lowering alcohol content in wine presents challenges due to legal constraints. Regulations typically permit the addition of water only for the preparation of enzymes, yeast, or other enological products, such as fining agents, not for diluting the sugar content of grape must [30]. Alternative techniques are being explored to achieve alcohol reduction, which could be grouped as viticultural or winemaking practices [31]. Viticultural practices could involve harvesting grapes at an early stage of maturity [32], double-pruning, to encourage bud growth during spring and summer, and shifting berry ripening to cooler periods of the season. This approach enhances the phenolic composition of the berries and ensures a more balanced sugar/phenolic content [33]. Introducing new grape varieties with delayed sugar accumulation and acid degradation [34], as well as implementing and adjusting cultural practices, such as post-veraison leaf removal [33]. In the winery, various enological techniques have been suggested to produce wines with lower alcohol content. For instance, utilizing yeast strains (*Saccharomyces* or non-*Saccharomyces*) with reduced ethanol production during winemaking [35] is one approach. Additionally, biotechnological solutions, mostly relying on the selection and improvement of yeast strains capable of metabolizing sugar into pyruvic acid, glycerol, and other compounds besides alcohol, are being investigated as a potential solution [36, 37]. Production of L-lactic acid by non-*Saccharomyces* as *Lachancea thermotolerans* through the enzymatic activity of lactate dehydrogenase (LDH) is an interesting approach to increase wine freshness and decrease the wine pH [38]. Alternative winemaking techniques to address the rising alcohol levels in red wines attributed to global warming have been developed, such as the practice of blending unripe grapes or must with fully ripened counterparts. Numerous studies involving various *Vitis vinifera* grape varieties, such as Shiraz, Cabernet Sauvignon, Pinot Noir, Malbec, Grenache, and Tannat, have explored the effects of blending unripe grapes or must with those from fully ripened grapes [32, 39–41]. These studies have demonstrated that blending unripe grapes or must with well-ripened counterparts results in wines with lower alcohol and pH levels and higher total acidity. Furthermore, the blending process positively impacts the phenolic composition, particularly tannins and anthocyanins, leading to enhancements in color and mouthfeel. Various physical methods are commonly employed in the removal of ethanol from finished wines, including membrane filtration, vacuum distillation, and supercritical fluid extraction [42, 43]. Among these, the spinning cone column and reverse osmosis system are frequently used [44, 45]. However, these methods require expensive equipment, and their impact on wine quality is still not well understood [42].

2.2 Adjustments in viticulture practices to reduce grapes temperatures

A wide array of techniques exists to delay grape ripening, which can be broadly categorized into two groups: alterations in viticulture practices and adjustments in plant material.

Shading, a technique used to mitigate high temperatures, has been examined in Sangiovese and Semillon grapevines [46, 47]. This method involves partially or completely covering grapevine canopies from budburst to harvest [48, 49]. The effects of this technique on canopy growth and berry development were studied, and it was found to decrease berry mass and increase the accumulation of phenolic substances, increasing total acidity [50].

Another mitigation strategy for climate change consists of the clonal selection of grape varieties as clonal variability, or genetic diversity within a grape variety, is a well-known phenomenon. Traditionally, clones have been chosen based on characteristics such as early ripening, high productivity, and elevated grapes sugar concentration. However, with the changing climate, new clones with contrasting characteristics may be more desirable. For example, sugar accumulation patterns differ significantly among clones, as demonstrated by a clonal selection trial with Cabernet Franc. The variation in grape sugar concentration among clones can be as high as 17 g/L (equivalent to 1% potential alcohol) at ripeness [51].

As temperatures continue to rise, traditional grape varieties may no longer ripen within the optimal period, potentially impacting the quality of wine produced. One potential strategy for adaptation is to shift toward planting later-ripening grape varieties. In European wine appellations, grape variety selection is usually regulated to ensure the best quality and typicity under local climatic conditions. However, as the climate changes, these regulations may need to be adjusted. A broader selection of grapevine varieties could serve as an important tool for adapting to climate change. As a result, grape varieties currently cultivated, especially those that ripen early, may no longer grow in their current locations under altered environmental conditions in the future [52]. This may be particularly relevant for regions already experiencing warm climates, where climate change may threaten the balanced ripening of grapes and the sustainability of existing varieties and wine styles [53, 54]. However, future warming in cooler climate regions may improve the suitability for producing high-quality wines [55].

It is crucial to explore mitigation alternatives to maintain grape quality [56]. One such technique involves the exogenous application of minerals, which can help reduce water usage while preserving or even enhancing wine quality. Silicon dioxide (SiO_2) particles, when applied as an aqueous suspension, form a physical or mechanical barrier (as precipitated amorphous silica) in the skin. Additionally, under conditions of water stress, the presence of Si may improve potassium uptake by plants [57]. Potassium plays a significant role in wine pH, which, in turn, affects chemical and microbiological stability, as well as the perception of wine flavor [58]. Numerous authors have concentrated their research on the advancement of eco-friendly practices, such as the application of kaolin, to maintain quality in a challenging climate. This requires the reduction of leaf and fruit berry surface temperature [59]. Indeed, the foliar application of solar protectants has already demonstrated promising results concerning the general fruit quality potential within the climate change context at a local scale [60, 61].

2.3 Vine pests and diseases associated with climate changes and their impact on wine quality

Climate change also poses a threat to grape production through the increased prevalence of pests and diseases along with the vectors that transmit these diseases. Grapes are susceptible to mycotoxigenic fungi, with *Aspergillus* being of particular concern due to its production of ochratoxin A (OTA). Mycotoxins are secondary metabolites produced by certain fungi and are concerning for human health due to their carcinogenic properties [62]. As a result, the levels of mycotoxins in certain food products intended for human consumption are strictly regulated, with specific maximum thresholds established. In the European Union, OTA is the only regulated mycotoxin for wine and other grape products, with a maximum allowable limit of 2 μg/kg [63].

The temperature ranges that result in high levels of OTA on grapes depend on the fungal species: *Aspergillus niger* aggregate strains and individual *Aspergillus carbonarius* have optimal ranges of 30–35°C and 25–30°C, respectively [64, 65]. The temperature's impact is most significant during infestation [66]. A temperature of 21°C appears to be the lower limit, below which fungal growth and OTA production are insufficient to reach critical levels in wine during the susceptible berry period [67]. However, it is widely recognized that climate change is leading to a redistribution of fungi, with a trend correlated with increasing temperatures and more frequent droughts in southern Europe [68]. Consequently, climate plays a crucial role in determining contamination levels once these fungi are established, with elevated temperatures being a key factor in the presence of OTA. Wines from warmer regions in southern Europe tend to have higher concentrations of OTA compared to those from cooler northern regions. Generally, OTA concentration in wine detected at 30°C exceeds that at 20°C in most cases [69]. Research suggests a correlation between mycotoxin occurrence and warm winemaking regions, likely to expand with global temperature increases. A review indicated such a correlation between climate and grape and wine OTA levels [70]. A survey of 942 wines by a regulatory laboratory found higher OTA concentrations in wines from southern European countries than from northern European countries [71]. It is suggested that as climate change advances, better-adapted species, such as *A. niger,* could thrive in southern Spain, potentially leading to a rise in fumonisins as OTA levels decrease [72]. The shift toward drier and hotter climates may favor the prevalence of *A. tubingensis* and *A. niger* over *A. carbonarius* as these species are better suited to extreme heat and aridity [73]. The current climate change scenario is expected to alter the geographical distribution of *Aspergillus flavus*, a concerning fungus. As higher temperatures may promote the dominance of more hazardous mycotoxins, such as aflatoxins, there is a possibility that they might surpass OTA as the main mycotoxin due to increased suitability for thermotolerant *Aspergilli* that produce aflatoxins [74, 75]. Recent reports have highlighted *A. flavus* as an emerging contaminant in vineyards, leading to discussions about the necessity of regulating aflatoxin levels in grapes. This widespread occurrence raises concerns about the potential contamination of grapes with aflatoxin.

These findings underscore the impact of climate change on the prevalence of mycotoxigenic fungi, such as *A. flavus* in vineyards, prompting a reevaluation of the risk associated with grape contamination by aflatoxin [76].

Establishing dependable models to evaluate the impact of climate change on vine health is essential, given its direct correlation with mycotoxin contamination. Although strides have been made in mitigating OTA levels in wine, the implications of climate change on other mycotoxins cannot be overlooked and will become increasingly critical [75].

3. Conclusions

Various strategies are being implemented at different stages of grape and wine production to tackle the significant challenges presented by rapid climate change in recent decades. To address rising temperatures, adjustments in viticulture practices and plant material are essential to ensure that actual harvest dates align with the optimal period. These adjustments involve selecting grapevine clones and adopting new approaches to manage vineyards, aiming to avoid grapes with elevated sugar levels, low acidity, and high pH, all of which significantly impact the sensory balance and microbiological stability of wine. In the winery, it will be necessary to adapt existing winemaking practices and develop novel ones. This includes incorporating a blend of unripe grapes with fully matured grapes, utilizing yeast strains with reduced ethanol production, and enhanced acid production during alcoholic fermentation. Moreover, implementing partial dealcoholization of alcoholic wines through gentle physical techniques, such as reverse osmosis, nanofiltration, spinning cone columns, and others, is essential. The impact of these techniques on wine quality should be thoroughly studied, including their effects on the wine aging process.

Acknowledgements

FCT-Portugal, CQ-VR (UIDB/00616/2020 and UIDP/00616/2020). (https://doi.org/10.54499/UIDP/00616/2020) and Project Vine&Wine Portugal—Driving Sustainable Growth Through Smart Innovation, Application n. ° C644866286-00000011, co-financed in the scope of the Mobilizing Agendas for Business Innovation, under Reg. (EU) 2021/241, in the Plano de Recuperação e Resiliência (PRR) to Portugal, na sua componente 5 - Capitalização e Inovação Empresarial.

Conflict of interest

The authors declare no conflict of interest.

Author details

Fernanda Cosme*, Luís Filipe-Ribeiro and Fernando M. Nunes
Chemistry Research Center—Vila Real (CQ-VR), Food and Wine Chemistry Laboratory, University of Trás-os-Montes and Alto Douro, Vila Real, Portugal

*Address all correspondence to: fcosme@utad.pt

References

[1] Vivier MA, Pretorius IS. Genetically tailored grapevines for the wine industry. Trends in Biotechnology. 2002;**20**(11):472-478. DOI: 10.1016/s0167-7799(02)02058-9

[2] Yang J, Xiao XY. Grape phytochemicals and associated health benefits. Critical Reviews in Food Science and Nutrition. 2013;**53**(11):1202-1225. DOI: 10.1080/10408398.2012.692408

[3] Zhu F, Du B, Zheng L, Li J. Advance on the bioactivity and potential applications of dietary fibre from grape pomace. Food Chemistry. 2015;**186**:207-212. DOI: 10.1016/j.foodchem.2014.07.057

[4] OIV State of the World Vitivinicultural Sector in 2022. International Organization of Vine and Wine. 2023. Available from: https://www.oiv.int/public/medias/7298/oiv-state-of-the-vitivinicultural-sector-in-2022.pdf [Accessed: February 17, 2024]

[5] Sweetman C, Sadras VO, Hancock RD, Soole KL, Ford C. Metabolic effects of elevated temperature on organic acid degradation in ripening *vitis vinifera* fruit. Journal of Experimental Botany. 2014;**65**:5975-5988. DOI: 10.1093/jxb/eru343

[6] Greer DH, Weston C. Heat stress affects flowering, berry growth, sugar accumulation and photosynthesis of *vitis vinifera* cv. semillon grapevines grown in a controlled environment. Functional Plant Biology. 2010;**37**:206-214. DOI: 10.1071/FP09209

[7] Kuhn N, Guan L, Dai ZW, Wu B-H, Lauvergeat V, Gomès E, et al. Berry ripening: Recently heard through the grapevine. Journal of Experimental Botany. 2014;**65**:4543-4559. DOI: 10.1093/jxb/ert395

[8] Caravia L, Collins C, Petrie PR, Tyerman SD. Application of shade treatments during Shiraz berry ripening to reduce the impact of high temperature. Australian Journal of Grape and Wine Research. 2016;**22**:422-437. DOI: 10.1111/ajgw.12248

[9] Lecourieux F, Kappel C, Pieri P, Charon J, Pillet J, Hilbert G, et al. Dissecting the biochemical and transcriptomic effects of a locally applied heat treatment on developing cabernet sauvignon grape berries. Frontiers in Plant Science. 2017;**53**:8. DOI: 10.3389/fpls.2017.00053

[10] Greer DH, Weedon MM. The impact of high temperatures on *Vitis vinifera* cv. Semillon grapevine performance and berry ripening. Frontiers in Plant Science. 2013;**4**:491. DOI: 10.3389/fpls.2013.00491

[11] Ovadia R, Oren-Shamir M, Kaplunov T, Zutahy Y, Lichter A, Lurie S. Effects of plant growth regulators and high temperature on colour development in "crimson seedless" grapes. Journal of Horticultural Science and Biotechnology. 2013;**88**:387-392. DOI: 10.1080/14620316.2013.11512980

[12] Martínez-Lüscher J, Sánchez-Díaz M, Delrot S, Aguirreolea J, Pascual I, Gomès E. Ultraviolet-B alleviates the uncoupling effect of elevated CO_2 and increased temperature on grape berry (*Vitis vinifera* cv. Tempranillo) anthocyanin and sugar accumulation. Australian Journal of Grape and Wine Research. 2016;**22**:87-95. DOI: 10.1111/ajgw.12213

[13] Lecourieux D, Kappel C, Claverol S, Pieri P, Feil R, Lunn J, et al. Proteomic and metabolomic profiling underlines the stage- and time-dependent effects of high temperature on grape berry metabolism. Journal of Integrative Plant Biology. 2020;**62**:1132-1158. DOI: 10.1111/jipb.12894

[14] Arrizabalaga-Arriazu M, Morales F, Irigoyen JJ, Hilbert G, Pascual I. Growth performance and carbon partitioning of grapevine Tempranillo clones under simulated climate change scenarios: Elevated CO_2 and temperature. Journal of Plant Physiology. 2020;**252**:153226. DOI: 10.1016/j.jplph.2020.153226

[15] Azuma A, Yakushiji H, Koshita Y, Kobayashi S. Flavonoid biosynthesis-related genes in grape skin are differentially regulated by temperature and light conditions. Planta. 2012;**236**:1067-1080. DOI: 10.1007/s00425-012-1650-x

[16] Gouot JC, Smith JP, Holzapfel BP, Walker AR, Barril C. Grape berry flavonoids: A review of their biochemical responses to high and extreme high temperatures. Journal of Experimental Botany. 2019;**70**:397-423. DOI: 10.1093/jxb/ery392

[17] Sadras VO, Moran MA. Elevated temperature decouples anthocyanins and sugars in berries of Shiraz and Cabernet franc. Australian Journal of Grape and Wine Research. 2012;**18**:115-122. DOI: 10.1111/j.1755-0238.2012.00180.x

[18] Spayd SE, Tarara JM, Mee DL, Ferguson JC. Separation of sunlight and temperature effects on the composition of Vitis vinifera cv. Merlot berries. American Journal of Enology and Viticulture. 2002;**53**:171-182. DOI: 10.5344/ajev.2002.53.3.171

[19] Sadras V, Petrie P, Moran M, Bastian S, Taylor D. Decompressing harvest and preserving wine style in warming climates. Australian and New Zealand Grapegrower & Winemaker. 2013;**594**:47

[20] Leolini L, Moriondo M, Romboli Y, Gardiman M, Costafreda-Aumedes S, de Cortázar-Atauri IG, et al. Modelling sugar and acid content in sangiovese grapes under future climates: An Italian case study. Climate Research. 2019;**78**:211-224. DOI: 10.3354/cr01571

[21] Gatti M, Garavani A, Cantatore A, Parisi MG, Bobeica NA, Merli MC, et al. Interactions of summer pruning techniques and vine performance in the white *vitis vinifera* cv. ortrugo. Australian Journal of Grape and Wine Research. 2015;**21**:80-89. DOI: 10.1111/ajgw.12107

[22] Mira de Orduña R. Climate change associated effects on grape and wine quality and production. Food Research International. 2010;**43**(7):1844-1855. DOI: 10.1016/j.foodres.2010.05.001

[23] Pons A, Allamy L, Schüttler A, Rauhut D, Thibon C, Darriet P. What is the expected impact of climate change on wine aroma compounds and their precursors in grape? OENO one. 2017;**51**(2):141-146. DOI: 10.20870/oeno-one.2017.51.2.1868

[24] Alston JM, Fuller KB, Lapsley JT, Soleas G. Too much of a good thing? Causes and consequences of increases in sugar content of California wine grapes. Journal of Wine Economics. 2011;**6**:135-159. DOI: 10.1017/S1931436100001565

[25] Drappier J, Thibon C, Rabot A, Gény L. Relationship between wine composition and temperature: Impact on Bordeaux wine typicity in the context of global warming. Critical Reviews in Food Science and Nutrition. 2019;**59**:14-30. DOI: 10.1080/10408398.2017.1355776

[26] Van Leeuwen C, Destrac-Irvine A, Dubernet M, Duchêne E, Gowdy M, Marguerit E, et al. An update on the impact of climate change in viticulture and potential adaptations. Agronomy. 2019;**9**:514. DOI: 10.3390/agronomy9090514

[27] Jones GV, Webb LB. Climate change, viticulture, and wine: Challenges and opportunities. Journal of Wine Research. 2010;**21**:103-106. DOI: 10.1080/09571264.2010.530091

[28] Jones GV. Climate, grapes and wine: Structure and suitability in a changing climate. Acta Horticulturae. 2012;**931**:19-28. DOI: 10.17660/ActaHortic.2012.931.1

[29] Saliba AJ, Ovington LA, Moran CC. Consumer demand for low-alcohol wine in an Australian sample. International Journal of Wine Research. 2013;**5**:1-8. DOI: 10.2147/IJWR.S41448

[30] Organisation Internationale de la vigne et du vin (OIV). International Code of Œnological Practices. 2019. Available from: http://www.oiv.int/public/medias/6558/code-2019-en.pdf

[31] Varela C, Dry P, Kutyna D, Francis I, Henschke P, Curtin C, et al. Strategies for reducing alcohol concentration in wine. Australian Journal of Grape and Wine Research. 2015;**21**:670-679. DOI: 10.1111/ajgw.12187

[32] Kontoudakis N, Esteruelas M, Fort F, Canals J, Zamora F. Use of unripe grapes harvested during cluster thinning as a method for reducing alcohol content and pH of wine. Australian Journal of Grape and Wine Research. 2011;**17**(2):230-238. DOI: 10.1111/j.1755-0238.2011.00142.x

[33] Martínez-Moreno A, Sanz F, Yeves A, Gil-Muñoz R, Martínez V, Intrigliolo D, et al. Forcing bud growth by double-pruning as a technique to improve grape composition of *Vitis vinifera* L. cv. Tempranillo in a semi-arid Mediterranean climate. Scientia Horticulturae. 2019;**256**:108614. DOI: 10.1016/j.scienta.2019.108614

[34] Schultz H. Climate change and viticulture: A European perspective on climatology, carbon dioxide and UV-B effects. Australian Journal of Grape and Wine Research. 2000;**6**(1):2 12. DOI: 10.1111/j.1755-0238.2000.tb00156.x

[35] Ciani M, Ferraro L. Enhanced glycerol content in wines made with immobilized Candida stellata cells. Applied and Environmental Microbiology. 1996;**62**(1):128 132. DOI: 10.1128/aem.62.1.128-132.1996

[36] Ciani M, Morales P, Comitini F, Tronchoni J, Canonico L, Curiel JA, et al. Non-conventional yeast species for lowering ethanol content of wines. Frontiers in Microbiology. 2016;**7**:642. DOI: 10.3389/fmicb.2016.00642

[37] Gonzalez R, Guindal AM, Tronchoni J, Morales P. Biotechnological approaches to lowering the ethanol yield during wine fermentation. Biomolecules. 2021;**11**(11):1569. DOI: 10.3390/biom11111569

[38] Hranilovic A, Albertin W, Capone DL, Gallo A, Grbin PR, Danner L, et al. Impact of Lachancea thermotolerans on chemical composition and sensory profiles of Merlot wines. Food Chemistry. 2021;**349**:129015. DOI: 10.1016/j.foodchem.2021.129015

[39] Schelezki OJ, Antalick G, Šuklje K, Jeffery DW. Pre-fermentation approaches to producing lower alcohol wines from cabernet sauvignon and shiraz: Implications for wine quality based on chemical and sensory analysis. Food Chemistry. 2020;**309**:125698. DOI: 10.1016/j.foodchem.2019.125698

[40] Piccardo D, Favre G, Pascual O, Canals JM, Zamora F, González-Neves G. Influence of the use of unripe grapes to reduce ethanol content and pH on the color, polyphenol and polysaccharide composition of conventional and hot macerated Pinot Noir and Tannat wines. European Food Research and Technology. 2019;**245**(6):1321 1335. DOI: 10.1007/s00217-019-03258-4

[41] Fanzone M, Sari SE, Mestre MV, Catania AA, Catelén MJ, Jofré VP, et al. Combination of pre-fermentative and fermentative strategies to produce Malbec wines of lower alcohol and pH, with high chemical and sensory quality. OENO one. 2020;**54**(4):1041-1058. DOI: 10.20870/oeno-one.2020.54.4.4018

[42] Gil M, Estevez S, Kontoudakis N, Fort F, Canals JM, Zamora F. Influence of partial dealcoholization by reverse osmosis on red wine composition and sensory characteristics. European Food Research and Technology. 2013;**237**:481-488. DOI: 10.1007/s00217-013-2018-6

[43] Schmidtke LM, Blackman JW, Agboola SO. Production technologies for reduced alcoholic wines. Journal of Food Science. 2012;**77**(1):R25-R41. DOI: 10.1111/j.1750-3841.2011.02448.x

[44] Pickering GJ. Low- and reduced-alcohol wine: A review. Journal of Wine Research. 2000;**11**(2):129 144. DOI: 10.1080/09571260020001575

[45] Lisanti MT, Gambuti A, Genovese A, Piombino P, Moio L. Partial dealcoholization of red wines by membrane contactor technique: Effect on sensory characteristics and volatile composition. Food and Bioprocess Technology. 2013;**6**:2289-2305. DOI: 10.1007/s11947-012-0942-2

[46] Cartechini A, Palliotti A. Effect of shading on vine morphology and productivity and leaf gas exchange characteristics in grapevines in the field. American Journal of Enology and Viticulture. 1995;**46**:227-234. DOI: 10.5344/ajev.1995.46.2.227

[47] Greer DH, Weston C, Weedon M. Shoot architecture, growth and development dynamics of Vitis vinifera cv. Semillon vines grown in an irrigated vineyard with and without shade covering. Functional Plant Biology. 2010;**37**:1061-1070. DOI: 10.1071/FP10101

[48] Joscelyne VL, Downey MO, Mazza M, Bastian SE. Partial shading of cabernet sauvignon and shiraz vines altered wine color and mouthfeel attributes, but increased exposure had little impact. Journal of Agricultural and Food Chemistry. 2007;**55**:10888-10896. DOI: 10.1021/jf0721621

[49] Rochfort S, Ezernieks V, Bastian SE, Downey MO. Sensory attributes of wine influenced by variety and berry shading discriminated by NMR metabolomics. Food Chemistry. 2010;**121**:1296-1304. DOI: 10.1016/j.foodchem.2010.01.067

[50] Rojas-Lara B, Morrison JC. Differential effects of shading fruit or foliage on the development and composition of grape berries. Vitis. 1989;**28**:27-63. DOI: 10.5073/vitis.1989.28.199-208

[51] Van Leeuwen C, Roby J-P, Alonso-Villaverde V, Gindro K. Impact of clonal variability in *Vitis vinifera* cabernet franc on grape composition, wine quality, leaf blade stilbene content and downy mildew resistance. Journal of Agricultural and Food Chemistry. 2013;**61**:19-24. DOI: 10.1021/jf304687c

[52] Wolkovich EM, de Cortázar-Atauri IG, Morales-Castilla I, Nicholas KA, Lacombe T. From Pinot

to Xinomavro in the world's future wine-growing regions. Nature Climate Change. 2018;**8**:29. DOI: 10.1038/s41558-017-0016-6

[53] Fraga H, Santos JA, Malheiro AC, Oliveira AA, Moutinho-Pereira J, Jones GV. Climatic suitability of Portuguese grapevine varieties and climate change adaptation. International Journal of Climatology. 2016;**36**:1-12. DOI: 10.1002/joc.4325

[54] Jones GV, Alves F. Impact of climate change on wine production: A global overview and regional assessment in the douro valley of Portugal. International Journal of Global Warming. 2012;**4**:383-406. DOI: 10.1504/IJGW.2012.049448

[55] Fraga H, Garcia de Cortazar-Atauri I, Malheiro AC, Moutinho-Pereira J, Santos JA. Viticulture in Portugal: A review of recent trends and climate change projections. OENO one. 2017;**51**:61-69. DOI: 10.20870/oeno-one.2017.51.2.1621

[56] Bernardo S, Dinis L-T, Machado N, Moutinho-Pereira J. Grapevine abiotic stress assessment and search for sustainable adaptation strategies in Mediterranean-like climates. A review. Agronomy for Sustainable Development. 2018;**38**:1-20. DOI: 10.1007/s13593-018-0544-0

[57] Pavlovic J, Kostic L, Bosnic P, Kirkby EA, Nikolic M. Interactions of silicon with essential and beneficial elements in plants. Frontiers in Plant Science. 2021;**12**:1224. DOI: 10.3389/fpls.2021.697592

[58] Mpelasoka BS, Schachtman DP, Treeby MT, Thomas MR. A review of potassium nutrition in grapevines with special emphasis on berry accumulation. Australian Journal of Grape and Wine Research. 2003;**9**(3):154-168. DOI: 10.1111/j.1755-0238.2003.tb00265.x

[59] Dinis LT, Ferreira H, Pinto G, BernardoS,CorreiaCM,Moutinho-PereiraJ. Kaolin-based, foliar reflective film protects photosystem II structure and function in grapevine leaves exposed to heat and high solar radiation. Photosynthetica. 2015;**54**(1):47-55. DOI: 10.1007/s11099-015-0156-8

[60] Coniberti A, Ferrari V, Dellacassa E, Boido E, Carrau F, Gepp V, et al. Kaolin over sun-exposed fruit affects berry temperature, must composition and wine sensory attributes of sauvignon blanc. European Journal of Agronomy. 2013;**50**:75-81. DOI: 10.1016/j.eja.2013.06.001

[61] Dinis LT, Bernardo S, Conde A, Pimentel D, Ferreira H, Félix L, et al. Kaolin exogenous application boosts antioxidant capacity and phenolic content in berries and leaves of grapevine under summer stress. Journal of Plant Physiology. 2016;**191**:45-53. DOI: 10.1016/j.jplph.2015.12.005

[62] Leong SL, Hocking AD, Pitt JI, Kazi BA, Emmett RW, Scott ES. Australian research on ochratoxigenic fungi and ochratoxin A. International Journal of Food Microbiology. 2006;**111**:S10-S17. DOI: 10.1016/j.ijfoodmicro.2006.02.005

[63] European Commission (EC). Commission regulation (EU) 2023/915 of 25 April 2023 on maximum levels for certain contaminants in food and repealing regulation (EC) No 1881/2006. Official Journal of the European Union. 2023;**119**:103-156

[64] Bellí N, Marın S, Sanchis V, Ramos AJ. Influence of water activity and temperature on growth of isolates of Aspergillus section Nigri obtained from grapes. International Journal of Food Microbiology. 2004;**96**:19-27. DOI: 10.1016/j.ijfoodmicro.2004.03.004

[65] Bellí N, Ramos AJ, Sanchis V, Marín S. Incubation time and water activity effects on ochratoxin A production by Aspergillus section Nigri strains isolated from grapes. Letters in Applied Microbiology. 2007;**38**:72-77. DOI: 10.1046/j.1472-765x.2003.01445.x

[66] Clouvel P, Bonvarlet L, Martinez A, Lagouarde P, Dieng I, Martin P. Wine contamination by ochratoxin A in relation to vine environment. International Journal of Food Microbiology. 2008;**123**:74-80. DOI: 10.1016/j.ijfoodmicro.2007.12.003

[67] Paterson RRM, Lima N. Further mycotoxin effects from climate change. Food Research International. 2011;**44**:2555-2566. DOI: 10.1016/j.foodres.2011.05.038

[68] Medina A, Akbar A, Baazeem A, Rodriguez A, Magan N. Climate change, food security and mycotoxins: Do we know enough? Fungal Biology Reviews. 2017;**31**:143-154. DOI: 10.1016/j.fbr.2017.04.002

[69] Anli E, Bayram M. Ochratoxin A in wines. Food Reviews International. 2009;**25**:214-232. DOI: 10.1080/87559120902956182

[70] Blesa J, Soriano JM, Molto JC, Manes J. Factors affecting the presence of ochratoxin A in wines. Critical Reviews in Food Science and Nutrition. 2006;**46**:473-478. DOI: 10.1080/10408390500215803

[71] Soleas GJ, Yan J, Goldberg DM. Assay of ochratoxin A in wine and beer by high-pressure liquid chromatography photodiode array and gas chromatography mass selective detection. Journal of Agricultural and Food Chemistry. 2001;**49**:2733-2740. DOI: 10.1021/jf0100651

[72] García-Cela E, Crespo-Sempere A, Gil-Serna J, Porqueres A, Marin S. Fungal diversity, incidence and mycotoxin contamination in grapes from two agro-climatic Spanish regions with emphasis on *Aspergillus* species. Journal of the Science of Food and Agriculture. 2015;**95**:1716-1729. DOI: 10.1002/jsfa.6876

[73] Garcia-Cela E, Crespo-Semprre A, Ramos AJ, Sanchis V, Marin S. Ecophysiological characterization of *Aspergillus carbonarius*, *Aspergillus tubingensis* and *Aspergillus niger* isolated from grapes in Spanish vineyards. International Journal of Food Microbiology. 2014;**173**:89-98. DOI: 10.1016/j.ijfoodmicro.2013.12.012

[74] Paterson RRM, Lima N. How will climate change affect mycotoxins in food? Food Research International. 2010;**43**:1902-1914. DOI: 10.1016/j.foodres.2009.07.010

[75] Paterson RRM, Lima N. Thermophilic fungi to dominate aflatoxigenic/mycotoxigenic fungi on food under global warming. International Journal of Environmental Research and Public Health. 2017;**14**:199. DOI: 10.3390/ijerph14020199

[76] Melguizo C, Patiño B, Ramos AJ, Vázquez C, Gil-Serna J. Reconsidering the co-occurrence of *Aspergillus flavus* in Spanish vineyards and aflatoxins in grapes. Agriculture. 2023;**13**:1998. DOI: 10.3390/agriculture13101998

Chapter 2

Review of Mycotoxins in Grapes and Grape Products

Fernanda Cosme, Miguel Ribeiro, Luís Filipe-Ribeiro and Fernando M. Nunes

Abstract

This review explores the presence of mycotoxins in grapes and grape products, focusing on various types such as ochratoxin A (OTA), aflatoxins, fumonisins, patulin, and others. The discussion encompasses multifaceted factors influencing mycotoxin occurrence, including environmental aspects, agricultural practices, and post-harvest handling. Advanced techniques for mycotoxin detection, such as chromatography and immunoassays, are explored, along with the challenges associated with these methods. Mitigation strategies, such as the implementation of good agricultural practices and good manufacturing practices, are presented. Additionally, emerging technologies for mycotoxin control are discussed, highlighting innovative approaches in the field. This overview aims to contribute to the complex realm of mycotoxins in grapes and grape products, offering a holistic understanding from detection to mitigation. The concluding remarks emphasize the significance of proactive measures to ensure the safety and quality of grape products regarding mycotoxin challenges.

Keywords: mycotoxins, grapes, grape products, mycotoxin detection, mitigation strategies

1. Introduction

Grape products encompass the grape berry and various products derived from grape berry processing, including wine, grape juice, distillates, vinegar, dried grapes, jams, and jellies. They can be susceptible to contamination with mycotoxins produced by specific fungi [1]. Literature data reveal the presence of various mycotoxin groups in grapes and nearly every type of commodity related to grapes [2–10]. Mycotoxins constitute a large and heterogeneous group of substances with diverse chemical structures and biological effects. They are low-molecular-weight secondary metabolites produced by filamentous fungi and can have toxic effects on humans, even at low concentrations. Mycotoxins are produced by a wide variety of fungal species, with the major fungal genera responsible for producing these substances being *Aspergillus*, *Penicillium*, *Fusarium*, and *Alternaria* [11, 12]. The genus *Penicillium* appears to be more prevalent in temperate and cold climates, such as those in northern Europe, whereas *Aspergillus* is more frequently associated with warmer and more humid regions [13].

IntechOpen

When it comes to fresh table grapes meant for direct consumption, numerous articles address mycotoxigenic fungi and mycotoxin contamination [14, 15]. Several studies have also examined mycotoxins and the associated fungi in dried grapes [16, 17]. Juices rank as the third-largest source of exposure to OTA following cereals and wines. Given that juices are often consumed by children, there is an added need for awareness concerning mycotoxin presence [6, 16–19]. Research studies are available for OTA [3, 20–25], being the subject of the majority of mycotoxin research related to grapes and grape products, aflatoxins [7, 16, 17, 26–28], fumonisins [4, 8, 20, 21, 29–31] patulin [12, 32, 33], citrinin [2, 34–36], and *Alternaria* toxins [11].

Many mycotoxins are highly stable metabolites that can withstand technological processes due to their resistance to heat, physical, and chemical treatments, thus remaining in the final products [37], except for patulin in wine, as it is degraded during the fermentation process [38]. Nevertheless, studies considering the effects of climate change on toxigenic fungi remain a challenge. Furthermore, the co-occurrence of OTA and other mycotoxins in grape products needs to be assessed to verify the risk of human exposure to these compounds.

2. Main types of mycotoxins found in grapes and grape products

Mycotoxins are considered among the most significant food contaminants due to their adverse impact on public health and food security. The molecular structures of mycotoxins vary widely, leading to a diverse range of effects on human health. The mycotoxins of greatest significance in grape and grape products include ochratoxin A, aflatoxins, fumonisin B2, and patulin [39].

2.1 Ochratoxin A

Ochratoxins constitute a group of toxic secondary metabolites produced by fungi. OTA stands out as the most toxic member of the ochratoxin group and has been extensively studied (**Figure 1**). It has been classified as a Group 2B carcinogen (i.e., a possible human carcinogen) by the International Agency for Research on Cancer (IARC) [40]. Nevertheless, OTA can be found in grape juice and wine due to grape contamination by toxigenic fungi [41]. The contamination of wine with OTA was initially reported in 1996 [42]. To address potential consumer risks linked to dietary exposure to OTA, the European Union has established maximum permitted levels for this mycotoxin in various grape products. These levels include 2 μg/kg in wine and grape juice, and 8 μg/kg in dried vine fruits (raisins, and sultanas) [43].

Dachery et al. [44] reviewed the occurrence of OTA in grape juice and wine and found that approximately 70% of samples may contain this toxin, with levels reaching up to ten times higher than the legal limits of 2 μg/L in the European Union [45].

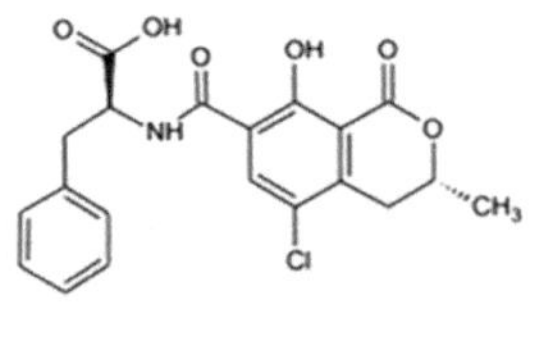

Figure 1.
Structure of ochratoxin A.

Currently, wine is recognized as the second most significant source of human exposure to OTA, following cereals. Major producers of OTA in grapes and dried vine fruits are *Aspergillus niger* var. *niger* and *Aspergillus carbonarius* [46].

Several authors have determined OTA levels in red, white, rose, Jerez, and sweet wines from different countries. In red wine, the values ranged from 0.58–2.36 μg/L and 0.05–0.35 μg/L in Italy [47, 48], 0.07–2.00 μg/L in Greece [49], 0.004–0.18 μg/L and 0.06–0.14 μg/L in Spain [50, 51], 0.01–0.24 μg/L in France [48], 0.41–0.45 μg/L and n.d.–0.17 μg/L in Portugal [52, 53], 0.39–7.96 μg/L, 0.04–1.92 μg/L, n.d.–0.82 μg/L, and 2.72–7.40 μg/L in Turkey [54–57], respectively; 0.02–4.82 μg/L in Argentina [58], 0.80–0.84 and 0.03–0.62 μg/L in Brazil [16, 59], 0.03–0.07 μg/L in Australia [48], 0.03–5.95 and n.d.–0.20 μg/L in China [60, 61], and 0.09–0.94 in Tunisia [62]. For white wine, the values ranged from 0.06–1.36 μg/L in Italy [47], 0.25–1.80 μg/L, 0.02–0.34 μg/L, n.d.–0.62 μg/L, and n.d.–4.41 μg/L in Turkey [54–57], respectively, n.d.–0.03 μg/L in Brazil [16, 17], 0.03–0.07 and n.d.–0.36 μg/L in China [60, 61], and from 0.11–1.50 μg/L in Tunisia [62]. In rose wines from Turkey, the values of OTA ranged from 0.03–2.23 μg/L and from n.d.–0.16 μg/L [54, 56]. In Jerez wine from Spain, the levels of OTA ranged from 0.04–0.64 μg/L [50]. In sweet wine from Italy, they ranged from 0.21–1.56 μg/L, in Greece from n.d.–2.82 μg/L, and in Spain from 0.01–4.63 μg/L [63–65], respectively. OTA contamination can occur at any stage of the winemaking process, from the early colonization of mycotoxigenic fungi in grapes to the final steps in the wine process. Nevertheless, the main source of contamination in the final product results from the transfer of mycotoxins from grapes [66]. Furthermore, the winemaking process significantly influences OTA content [67–69], with higher concentrations reported in red wines (<0.01–7.63 μg/L), compared to rose (<0.01–2.40 μg/L) and white wines (<0.01–1.72 μg/L) in general [64, 70–72]. In a survey regarding the presence of OTA influenced by the type of wine, it was found that 25% of white wines, 40% of rosé, and 54% of red wines were contaminated with OTA [73]. For red wines, the maceration process can lead to an increase in OTA content (approximately 20%) [9] due to the extended contact between grape skins and grape juice, facilitating the solubility and diffusion of this mycotoxin from contaminated skins [74]. Conversely, the absence of maceration in white and rose wines appears to be a critical factor contributing to low OTA levels in these wines [6].

2.2 Aflatoxins

There are approximately 20 types of aflatoxins; however, the four major aflatoxins found in foods are aflatoxin B_1 (AFB_1), aflatoxin B_2 (AFB_2), aflatoxin G_1 (AFG_1), and aflatoxin G_2 (AFG_2), identified based on their fluorescence under UV light (blue or green) and relative chromatographic mobility during thin-layer chromatography (**Figure 2**). Their chemical structure is very similar, classified as furanocoumarins due to the presence of a coumarin nucleus associated with a furan and lactone ring. AFB_1 is the most potent natural carcinogen known and is typically the major aflatoxin produced by toxigenic strains, has been classified as a human carcinogen (Group 1) by the International Agency for Research on Cancer (IARC) [40]. Aflatoxins are produced through a polyketide pathway by numerous strains of *Aspergillus flavus* and *Aspergillus parasiticus* [75]. Notably, *Aspergillus flavus* is a prevalent contaminant in agriculture. Despite this threat, studies conducted in certain Mediterranean countries have reported a low incidence of *Aspergillus flavus* in vineyards and AFB_1 in grapes and derived products [76]. However, El Khoury et al. [7] observed an increasing incidence of vineyards contaminated by *Aspergillus flavus* isolates and grapes containing

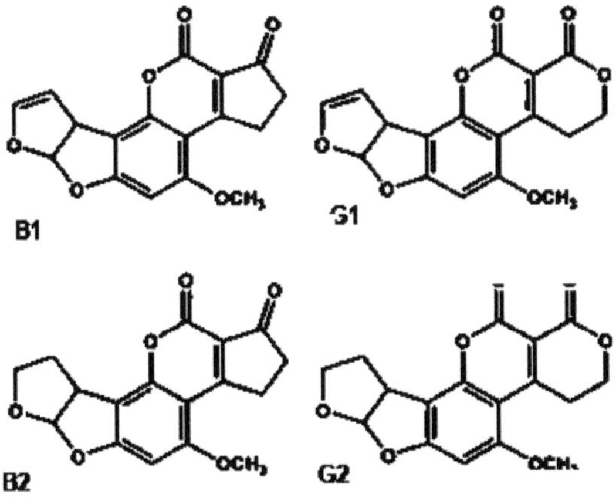

Figure 2.
Structure of aflatoxins B1, B2, G1, and G2.

AFB_1. Aflatoxins and aflatoxin-producing strains have been detected in grapes and grape musts in North African countries—Lebanon, Morocco, and Tunisia—and in Asia—China [7, 77–80]. AFB_1 was detected in 40% of grape musts from twenty-seven vineyards in Lebanon, with levels below 0.46 μg/L [7]. In Tunisian vineyards, *Aspergillus flavus* constituted 23% of mycotoxin-producing fungi in grapes, generating AFB_1 at concentrations ranging from 21 to 54 μg/g of the culture medium [78]. The current European Commission Regulation (EU) No 915/2023 sets maximum residue limits (MRLs) total aflatoxins (sum of B_1, B_2, G_1, and G_2) to 4 μg/kg and AFB_1 to 2 μg/kg in dried fruit intended for direct human consumption. For dried fruits intended for sorting or other physical treatments before consumption or inclusion as an ingredient in food products, the MRLs are set at 10 μg/kg for total aflatoxins (sum of B_1, B_2, G_1, and G_2) and 5 μg/kg for AFB_1 [43]. Therefore, it is crucial to comprehend the level of contamination with toxin-producing fungi to prevent the introduction of toxin contamination into the food chain. The occurrence of aflatoxins in wines has been reported in recent years [63, 81]. AFB_1 was detected in red wine from Spain in the range of 1.25–13.43 μg/L [81]. In sweet wine from Italy, Di Stefano et al. [63] found AFB1 ranging from 0.017–0.035 μg/L, and AFB2 from 0.013–0.016 μg/L. A screening in different wines in Romania measured AFB1 ranging from 12.12 to 33.68 μg/L [82]. However, there is no established limit for aflatoxins in wines, and studies on the content of aflatoxins in wines are limited. While OTA currently remains the most prevalent mycotoxin in grapes, recent studies suggest that in certain European regions, AFB1 may eventually replace OTA as the primary concern due to climatic conditions, especially higher temperatures, which directly impact fungus development [1, 83].

2.3 Fumonisins

Fumonisins were initially described and characterized in 1988 (**Figure 3**) [84, 85]. The fungus *Aspergillus niger*, involved in the production of fumonisin B2, has been found on grape skin surfaces [86]. They are believed to be synthesized through the condensation of the amino acid alanine into an acetate-derived precursor. Fumonisin B2 has already been detected in grapes, raisins, grape must, and wine [10, 87]. The amount of fumonisin B2 produced by aspergilli in grapes appears to be low (0.1–7.8 mg/kg) [10, 88]. Fumonisin B2 was found in 9 out of 45 red wine samples

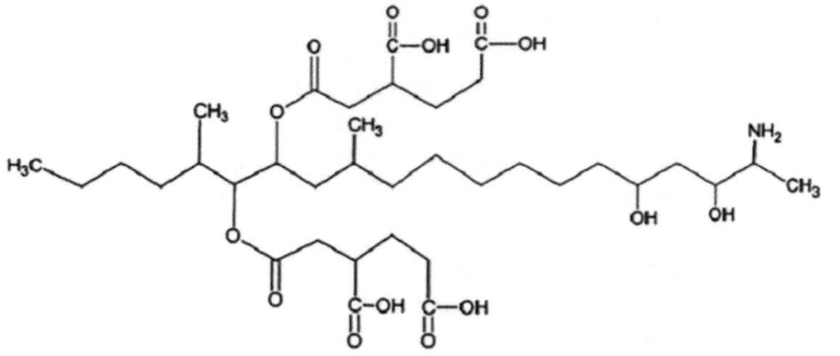

Figure 3.
Structure of fumonisin B2.

from Italy in the range of 0.4–2.4 μg/mL [89]. In a study performed by Mogensen et al. [10] on 77 wines from 13 different countries, the range of fumonisin B2 was from 1 to 20 μg/L. Also, Tamura et al. [90] found fumonisin B2 in both red and white wines.

2.4 Other mycotoxins

Patulin (**Figure 4**) is a water-soluble lactone produced via the polyketide metabolic pathway by many species, such as those within the *Penicillium* and *Aspergillus* genera, predominantly *Aspergillus clavatus*, *Penicillium expansum*, *Penicillium clavigerum*, and *Byssochlamys fulva* [91]. The Codex Alimentarius Commission [92] has recommended a maximum level of 50 μg/L for fruit juices and their products. In the case of fruit-based baby food, the European Union has strict legislation limiting the occurrence of patulin to a concentration of <10 μg/kg [43]. The results from Ostry et al. [34] showed that the occurrence of patulin and citrinin in the 23 grape must samples was 10 (43%) samples for patulin (range: 143–644 ng/g) and 2 (9%) samples for citrinin (range 2.5–3.5 ng/g).

Alternaria toxins (**Figure 4**) were detected in red and white wines in Argentina in the range of 13–18 μg/L by Broggi et al. [93]. In the Netherlands, in red wines, the range was 2–11 μg/L [94], and in Germany, the level ranged from 65 to 7.65 μg/L [95].

3. Factors influencing mycotoxin presence in grapes and grape products

It has been demonstrated that environmental stress conditions, such as drought, insect infestation, mechanical damage, cultivar susceptibility, nutritional

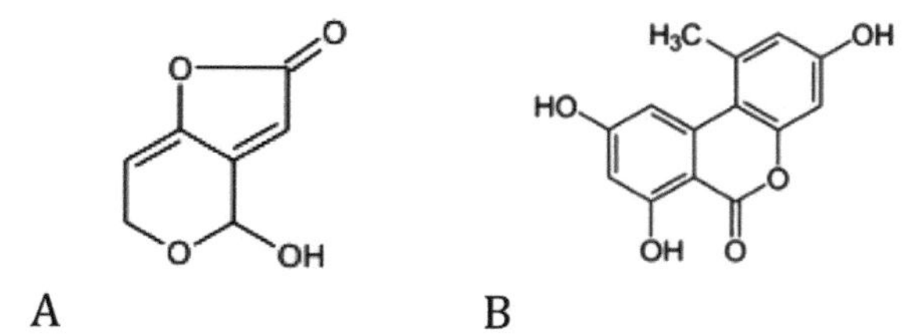

Figure 4.
Structure of patulin (A) and Alternaria toxins (B).

deficiencies, and unusual temperature, rainfall, or humidity, promote the dissemination of fungal populations, including those present on grapes and grape products. Indeed, alterations in farming practices may lead to increased stress on plants, consequently promoting fungal invasion and mycotoxin contamination [96]. The primary source of fungi in vineyards is soil [97]. Fungi are commonly found in the soil and on grape berries throughout the entire crop production cycle, although they face difficulty in penetrating healthy berries during early grape growth stages. Berries may develop microfissures and soften with ripening, thereby increasing nutrient availability and enabling further contamination by fungi. Their entry into the fruit is also facilitated by skin damage caused by insects, birds, or other factors such as rainfall or fungal infections [98]. Among these factors, insect pests have been confirmed to significantly contribute to fungal invasion of berries and their contamination with OTA. However, despite the numerous insect species that infest grapes [99], only a limited number of them, such as the European grapevine moth *Lobesia botrana* or mealybugs, have been linked to OTA contamination [100, 101]. The humidity in the vineyard may promote fungal proliferation, posing a significant concern [83]. Various factors, including weather conditions, harvest timing, biotic factors, agricultural and harvesting practices, or the type of wine, may also impact the final concentration of mycotoxins [102]. Importantly, winemaking practices, like maceration, can increase mycotoxin content in wines, and consequently, it is frequently detected in red wines, followed by rose and white wines. Additionally, Jiang et al. [80] suggested that the grape cultivar, especially its skin thickness, may affect the OTA levels found in wine. While numerous studies have explored factors influencing the occurrence of OTA in wine, there is a scarcity of research on aflatoxins, patulin, alternariol, and fumonisin B2 [1]. Additionally, it is possible that other mycotoxins may become more prevalent with the progression of climate change [103]. Therefore, there is a clear need for more information on the impact of climate change scenarios and environmental conditions such as temperature, water activity, and humidity on the fungal colonization, growth, and mycotoxin production by key mycotoxigenic species in the genera *Aspergillus*, *Fusarium*, *Penicillium*, and *Alternaria* [104, 105]. Aflatoxins may increase in temperate regions due to rising temperatures. For instance, in Italy during 2003 and 2004, hot and dry conditions led to *Aspergillus flavus* colonization and aflatoxin production [106].

4. Detection and analysis of mycotoxins from grapes and grape products

Detection and analysis of mycotoxins from grapes and grape products are imperative for ensuring food safety. However, challenges in mycotoxin analysis exist, including the discovery of new mycotoxins without a fully understood toxicity profile, modifications that can hamper their identification, co-occurrence, and the complexity of matrices [107]. Analytical approaches, including sample preparation methods, have evolved to overcome some of these limitations, covering a wider range of mycotoxins. Two major trends in mycotoxin analysis can be identified. Firstly, methods based on chromatography, with high sensitivity and specificity, facilitate the separation and quantification of mycotoxins, especially when combined with mass spectrometry. The other trend is immunochemical-based methods, such as the Enzyme-Linked Immunosorbent Assay (ELISA), which offer rapid and cost-effective analyses [107, 108].

4.1 Sampling, sample preparation, and extraction

The sampling, preparation, and extraction of mycotoxins are crucial steps for ensuring precision and accuracy in analytical procedures, especially in solid samples where their occurrence is known to be highly heterogeneous. The sampling process is essential in the analysis of both whole and dried grapes. In this context, in addition to adequate sampling, homogenization of the sample is necessary, for example by grinding [109]. In the EU, there are regulations for the sampling method for the official control of OTA levels in dried grapes, grape juice, grape must, and wine [43]. However, in the case of wine, grape juice, and grape must, sampling is simpler due to the liquid nature of these products. An extraction and preconcentration step are often required before their analysis by chromatographic techniques. The selection of the extraction solvent depends on the physical and chemical characteristics of the analyte, solvent cost, and safety; organic solvents such as acetonitrile, methanol, acetone, ethyl acetate, and dichloromethane are generally used for mycotoxin extraction. In liquid matrices, it is common to use a liquid-liquid extraction (LLE) approach followed by a cleaning step using solid phase extraction (SPE) or immunoaffinity columns (IAC) [63]. In contrast, for solid samples such as dried grapes, the solid-liquid extraction (SLE) procedure is preferred before the cleaning step. It should be noted that highly sensitive and specific immunoassays, such as ELISA, may not necessitate a clean-up step. On the other hand, techniques with poor resolution, like thin-layer chromatography (TLC), will likely require thorough clean-up procedures. Newer techniques such as QuEChERS (which stands for Quick, Easy, Cheap, Effective, Rugged, and Safe) and dispersive liquid-liquid microextraction (DLLME) have also been successfully used for detecting important mycotoxins in wines [110, 111].

4.2 Chromatography-based methods

Mycotoxin separation is typically achieved through chromatographic methods, including thin-layer chromatography (TLC), gas chromatography (GC), or liquid chromatography (LC). Due to the non-volatile nature of many mycotoxins in grapes and grape products, such as OTA, samples are generally separated by TLC or LC and quantified using fluorescence or mass spectrometry (MS) [112]. TLC has become a largely neglected technique, although it is still used in some laboratories, especially in developing countries, and when combined with an ultraviolet (UV) or fluorescence scanner. It is an easy-to-use, fast, and highly versatile separation technique, particularly important in resource-limited settings. Unlike more complex techniques, TLC offers rapid results and requires minimal specialized equipment, making it accessible for various applications. Notably, TLC has proven important in detecting and quantifying aflatoxins at low levels. Nevertheless, it has been superseded by more advanced methods such as GC and high-performance liquid chromatography (HPLC). Regarding GC, which was previously widely used for multiple analyses of trichothecene, modern advancements in HPLC with UV and fluorescence detectors (FLD), and more recently, in MS, have effectively supplanted it. As a result, HPLC, namely reversed-phase HPLC (RP-HPLC), has become the main method for determining mycotoxins and remains so in most laboratories (**Table 1**) [108].

Chromatographic techniques coupled with fluorescence detection are mainly used for confirmatory analyses, especially in the quantification of OTA. However, an additional confirmation step may be necessary to guarantee the identity of OTA. This confirmation is normally performed through derivatization with BF_3 in methanol or

Sample	Sample preparation	Mycotoxin	Method	Reference/ Year
Grape	QuEChERS	Aflatoxin ($B_1/B_2/G_1/G_2$), Cytochalasin (B, C, D), Diacetoxyscirpenol, Deoxynivalenol, 3-acetyl deoxynivalenol, 15-acetyl deoxynivalenol, Enniatin (A, A1, B, B1), Fusarenone X, HT-2/T-2 toxins, Neosolaniol, Ochratoxin (A, B, C)	UHPLC-HRMS	[110]/2024
Wine	LLE	Aflatoxin (B_1, B_2, G_1, G_2), Ochratoxin A, Zearalenone	UHPLC–MS/MS	[113]/2022
Wine	SPE	Ochratoxin (A, B)	HPLC-FLD	[114]/2020
Wine	Dilution and filtration	Aflatoxin (B_1, B_2, G_1, G_2), Deoxynivalenol, Diacetoxyscirpenol, Fumonisin (B1, B2), HT-2/T-2 toxins, Ochratoxin A, Zearalenone; pesticides (n = 185)	UHPLC–MS/MS	[115]/2019
Grape	QuEChERS	Alternariol, Alternariol monomethyl ether, Altenusin, Altenuene, Tentoxin, Tenuazonic acid	UHPLC–MS/MS	[11]/2019
Grape juice, wine	SPE	Ochratoxin A	LC-FLD	[116]/2018
Wine	IAC	Ochratoxin (A, B), Aflatoxin (B_1, B_2, G_1, G_2)	HPLC-FLD	[63]/2015

Table 1.
Examples of recent works on chromatography-based methods for mycotoxin determination in grapes and grape products.

with methanol in concentrated hydrochloric acid, both producing the methyl ester derivative of OTA [109].

More recently, ultra-high-performance liquid chromatography (UHPLC) coupled with high resolution mass spectrometry (HRMS) is considered the most advanced technique for the qualitative and quantitative analysis of mycotoxins, particularly for multiclass determination of mycotoxins, even in the presence of several interfering substances in complex samples. In fact, this state-of-the-art technique responds to current demands for multi-target techniques that guarantee specificity and sensitivity characteristics, while simultaneously allowing the identification of non-target compounds [107, 110]. This is of significant interest within the context of climate change. Anticipated shifts in climate variables, including temperature, rainfall, and atmospheric carbon dioxide levels, are expected to exert an influence on fungal growth and toxin production. The adoption of comprehensive analyses, rather than a focused approach, holds the potential to provide critical insights as these climate-induced changes progress [108, 117].

4.3 Immunochemical-based methods

Immunochemical-based methods serve as rapid initial screening tools, often provided in purpose-tailored kits. Leveraging the binding affinity between mycotoxins and specific antibodies, these methods facilitate rapid, *in situ* detection of prevalent mycotoxins in grapes and grape products. Despite their value in routine monitoring, they have limitations, such as the detection of many targets per sample and the inability to detect mycotoxins that are not included in the assay or those that have undergone chemical modifications. Nevertheless, they play a crucial role in rapid assessments, proving particularly beneficial in applications such as food safety and agricultural practices [112].

A widely used immunochemical method is ELISA. Commercially available ELISA-based kits for all regulated mycotoxins ensure food safety throughout the supply chain. ELISA, with its high sensitivity and specificity, is suitable for the quantitative measurement of various mycotoxins. Predominantly, classical and competitive inhibition formats are used, which perform better considering the limited epitope site display of mycotoxins [107].

Beyond ELISA, ongoing developments in immunochemical-based tests aim to create rapid, portable, and easy-to-operate systems. Among these, lateral flow immunoassays (LFIA), also known as lateral flow tests or immunochromatographic assays, are often used in on-site mycotoxin screening for grapes. LFIA provides fast results, easy operation, and requires minimal equipment. It involves moving the sample along a strip containing immobilized antibodies, generating a visible signal indicating the presence or absence of the target mycotoxin. LFIA, recognized for its simplicity and speed, is valuable for qualitative assessments, especially in scenarios requiring immediate results. However, its quantitative precision is limited [112, 118].

In the field of research, and with more limited market availability and commercial applications, biosensors use the selectivity and affinity of antibodies coupled to different sensor devices to determine prevalent mycotoxins. Biosensors based on synthetic ligands, designed to mimic the binding capacity of natural antibodies, also contribute to this growing field of research [119]. In general, biosensors come in various types and can face challenges including complex construction, costly labeling markers, and susceptibility to factors like temperature, pH, and immobilization support, which can affect sensitivity. Issues like cross-reactivity with similar compounds in the sample matrix, such as wine, can further complicate biosensor measurements, impacting their reliability. While biosensors hold promise, their broad practical application is still in development [120].

5. Mitigation strategies for mycotoxins in grapes and grape products

5.1 Viticultural practices to minimize mycotoxins in grapes and grape products

There are many fungi on grapes in the vines, but the main problem from the viewpoint of mycotoxin contamination is the black *Aspergilli*, specifically *Aspergillus carbonarius, A. niger, and A. welwitschiae* [86, 121, 122]. These types of fungi can produce OTA, contaminating the grapes and respective grape products like wine, grape must, and dried grapes. High temperature, humidity, and the region of cultivation are crucial factors that influence OTA accumulation in grape berries. The intensity of *Aspergillus* rot is highly influenced by excessive irrigation and rainfall before the harvest, causing berry splitting. Additionally, wounds on berries generated by insect attacks allow preferential entry points for black *Aspergilli*. High OTA levels can occur in grapes gravely damaged by the grape moth, *Lobesia botrana*, especially in Mediterranean regions [121].

However, diverse practices and good management of cultivation, pruning, and adequate irrigation can reduce the levels of black *Aspergilli* in the soil, consequently minimizing contaminations of grapes by these fungi. There are several approaches to minimize damage to grapes, such as open vineyard canopies, grape varieties with resistant skin to avoid rain damage, and the control of insect attacks and fungal diseases (e.g., *Botrytis* bunch rot, mildew). These practices can decrease the incidence of *Aspergillus* rot in mature berries. The problem of OTA in table grapes can be significantly reduced by attentive visual inspection to avoid damaged and discolored berries. Storing table grapes at low temperatures (0°C) with the application of some sulfur dioxide can reduce the severity of attacks by black *Aspergillus* species on the grapes. Wine grapes present a lower risk of OTA contamination than dried vine fruit because the ratio of *A. carbonarius* to *A. niger* increases during the drying process [123]. The dominant species in vineyards at harvest is *A. niger*, which rarely produces OTA [124]. However, the grape drying process could increase the survival of *A. carbonarius*, and almost all strains of this species can produce OTA, frequently to a water activity of 0.92 [125]. In conclusion, for dried grapes production, it is important to avoid berry damage and implement a rapid drying process. Grape cleaning and grape selection are important to remove dark berries and reduce OTA levels in the final products [86, 121].

5.2 Strategies and treatments to mitigate the final content of mycotoxins in wine

An important mycotoxin frequently found in wines is OTA, which is limited to a maximum of 2.0 μg/L in the European Union. In winemaking, harvesting healthy grapes, followed by a rapid process and maintaining good clean conditions in the winery, is crucial to minimize OTA wine contamination. Many vinification operations, such as pressing grapes and clarification processes that remove solids in suspension, some grape proteins, and spent yeast, help eliminate a significant quantity of OTA [86, 121]. Therefore, prevention and avoiding its appearance in wines are essential; when present, more efficient treatments may be necessary to remove OTA and ensure safe levels for human consumption [14].

Thus, knowledge about OTA contamination and its appearance in wine and wine by-products is crucial to manage OTA risk in contaminated stock. In some winemaking experiences, only 4% of the OTA present in grapes continues into the wine; the majority is eliminated in the pressed grape pomace. However, OTA content remains constant in wine even after 1-year of aging and is also present in all liquid fractions collected during vinification (i.e., grape juice, free-run juice, and wine after first and second sedimentation) [121]. Various physical, chemical, and microbiological methods have been described for OTA elimination from food products [14, 66, 68, 126–128].

Biological methods utilize microorganisms such as yeast, bacteria, and fungi for the detoxification of mycotoxins [66, 129, 130]. Bacteria, such as *Lactobacillus plantarum* and *Oenococcus oeni*, could adsorb OTA, with polysaccharides and peptidoglycan presumably involved in toxin binding [131]. Del Prete et al. [131] investigated 15 strains of enological lactic acid bacteria to verify their in vitro capacity to remove OTA and observed *Oenococcus oeni* as the most effective, with OTA reductions of 28%. According to Piotrowska [132], thermally inactivated bacterial biomass has a higher binding capacity for OTA (46.2–59.8%). Another study shows that OTA biodegradation by *Pediococcus parvulus* UTAD 473 can utilize 100% of OTA (1000 μg/L) within 7 days at 30°C in MRS broth medium, and 80% of OTA (7 μg/L) after 6 days of incubation in grape juice. However, no obvious degradation of OTA (7 μg/L) was observed in synthetic wine, perhaps due to the presence of ethanol, which could inhibit the enzyme responsible for the OTA degradation [126].

There are some doubts in the literature about *Saccharomyces cerevisiae* biodegradation capacity because almost all works demonstrate only its adsorption capacity [66].

S. cerevisiae was responsible for biodegrading 41% of 0.3 mg OTA/L after 24 h at 30°C, but details were not provided about the mechanism involved [133]. During must fermentation, yeasts adsorbed a maximum of 21% of the added OTA [134]. Also, almost 30% of the added OTA was removed after extended contact with yeast biomass [135]. However, Csutorás et al. [136] also verified that OTA (4000 μg/mL) could be adsorbed by *S. cerevisiae* along the winemaking process (90 days) in 73, 85, and 90% in white, rosé, and red wine grape juices, respectively. Finally, Petruzzi et al. [137] conducted a review on OTA removal primarily by yeast products: live cells, cell walls, cell wall extracts, and yeast lees, and concluded that yeast biomass could be considered a good treatment due to its adsorbing capacity, attributed to specific macromolecules in the cell wall, such as β-glucans and mannoproteins.

Some physical treatments have shown efficacy in OTA removal, such as wine filtration through a 0.45 μm membrane, which can reduce about 80% of OTA in the wine [138]. Solfrizzo et al. [139] found that OTA levels in wine can be significantly reduced by the repassage of grape pomaces through the wine. They also observed that if the grape pomaces are from the same grape variety as the treated wine, there will not be any significant impact on wine quality; however, it is not a practical treatment due to logistics difficulties in the winery. Other physical treatments like pulsed light (PL) can also efficiently remove OTA from clarified grape juice. The factors of OTA degradation by PL were explored, and the OTA removal presents a performance of 95.29% after optimization by response surface methodology [140]. Also, radiation methods including ultraviolet radiation, gamma radiation, electron beam radiation, and X-ray radiation can be safe and efficient in removing mycotoxins from food [141]. Inorganic mineral adsorbents present an efficient means of OTA removal due to their large specific surface area and ion adsorption capacity, primarily aluminum silicate, hydrated sodium calcium aluminosilicate, bentonite, zeolite, diatomaceous earth [142, 143]. Some studies have been conducted, especially with activated carbon. Activated carbon can effectively reduce OTA levels in wine; however, it may have a negative impact on wine quality [144]. Activated carbon is used as a fining agent in enology [145] at the maximum dosage of 1 g/L. The adsorption capacity of activated carbon is strictly dependent on its physicochemical characteristics, especially, pore structure and size, magnitude, pore volume, and distribution on the surface area, which are crucial for its adsorption capacity and performance [146].

Galvano et al. [147] verified that activated carbon demonstrates effective absorption of OTA in a model solution (1 milligram of activated carbon can remove 125 μg of OTA). However, the complexity of the wine matrix, particularly in red wines, could significantly reduce the OTA adsorption capacity of activated carbons due to the presence of many phenolic compounds. Seven commercially available deodorizing activated carbons were tested to optimize OTA removal from white and red wines at high levels of OTA (10.0 μg/L) [144]. The study showed that 1 g/L of activated carbon can remove 100% of OTA in white wines. It was also observed that activated carbon efficiency was less dependent on activated carbon physicochemical characteristics. However, in red wines, only one activated carbon tested could remove 100%, whereas the activated carbon with a higher abundance of mesopores exhibited better performance in OTA removal. The lower efficiency in red wines was associated with the competition of red wine anthocyanins for activated carbons mesopores [143].

AFB_1 and AFB_2 are two highly toxic mycotoxins that have been sometimes found in wines. For example, AFB_1 was present in 23% of the analyzed wines, and 21, 40, and 37% of the wines also contained AFG_1, AFB_2, and AFG_2, respectively. The concentration of these aflatoxins averaged 0.025, 0.043, 0.015, and 0.027 μg/L for AFB_1, AFG_1,

AFB_2, and AFG_2, respectively [148]. Previous studies have shown that the processing applied to grape products may influence aflatoxin levels. For instance, Heshmati et al. [149] observed a significant effect on aflatoxins removal during pekmez preparation, resulting in a significant reduction of AFB_1, AFB_2, AFG_1, and AFG_2 by up to 60.4, 76.7, 76.3, and 86.7%, respectively [149]. Interesting results were also obtained [150] during vinegar production from grapes, where they found that the values of AFB_1, AFB_2, AFG_1, and AFG_2 were significantly reduced by 76.20, 71.06, 69.26, and 75.85%, respectively. Alcoholic fermentation had the most significant effect on aflatoxin reduction during vinegar production, decreasing AFB_1, AFB_2, AFG_1, and AFG_2 by 41.87, 45.34, 45.37, and 46.52%, respectively. Currently, no authorized treatmenta are available for removing aflatoxins from wines. Several authorized fining agents like potassium caseinate, chitosan, bentonite, and activated carbon were tested for aflatoxins removal (AFB_1 and AFB_2). Interestingly, the work showed that the fining agents' performance in aflatoxins removal was dependent on the wine matrix, more so in white wine than in red wine. The most efficient fining agent was bentonite, which could remove 10 μg/L from white wine (100% of AFB_1) and 82% of AFB_2 from red wine. The impact of bentonite on white wine chromatic characteristics was low (color difference, $\Delta E^* = 1.35$). However, the treatment with bentonite could have a significant impact on red wine chromatic characteristics ($\Delta E^* = 4.80$) due to the considerable reduction of total anthocyanins, decreasing the respective wine color intensity by 1.5 points [151].

6. Conclusions

The presence of fungal contamination in vineyards can elevate health risks attributed to mycotoxins in grapes and their respective products. Among them, OTA is the most frequently reported mycotoxin in grapes and grape products, and this compound is regulated in several countries. OTA is initially removed from grape juice by its adsorption onto solid parts of grapes by yeasts and bacteria responsible for alcoholic or malolactic fermentation, which are usually removed from the wine during stabilization and bottling. When these mycotoxins persiste in the final wine, they can be efficiently removed by fining, for example, activated carbons. However, some studies have indicated the co-occurrence of other mycotoxins such as aflatoxins, fumonisin, and patulin, indicating that these metabolites cannot be disregarded. The presence of patulin is primarily a preharvest issue in table grapes and grape juice. More studies focusing on mycotoxin presence are needed, as well as research on the possible effects of processing grapes and grapes products on mycotoxin levels, which may become a future concern for the grape and grape products industries. It is also important to consider the scenario of climate change, as fungal development in grapes and grape products may deviate from the current situation in each region worldwide.

Acknowledgements

FCT-Portugal, CQ-VR (UIDB/00616/2020 and UIDP/00616/2020). (https://doi.org/10.54499/UIDP/00616/2020 and https://doi.org/10.54499/UIDB/00616/2020) and Project Vine&Wine Portugal—Driving Sustainable Growth Through Smart Innovation, Application n. ° C644866286 00000011, co-financed in the scope of the

Mobilizing Agendas for Business Innovation, under Reg. (EU) 2021/241, in the Plano de Recuperação e Resiliência (PRR) to Portugal, na sua componente 5—Capitalização e Inovação Empresarial.

Conflict of interest

The authors declare no conflict of interest.

Author details

Fernanda Cosme*, Miguel Ribeiro, Luís Filipe-Ribeiro and Fernando M. Nunes
Chemistry Research Centre—Vila Real, (CQ-VR), Food and Wine Chemistry Laboratory, University of Trás-os-Montes and Alto Douro, Vila Real, Portugal

*Address all correspondence to: fcosme@utad.pt

References

[1] Russell R, Paterson RRM, Armando V, Lima N, Guilloux-Bénatier M, Rousseaux S. Predominant mycotoxins, mycotoxigenic fungi and climate change related to wine. Food Research International. 2017;**103**:478-491. DOI: 10.1016/j.foodres.2017.09.080

[2] Abrunhosa L, Paterson RRM, Kozakiewicz Z, Lima N, Venâncio A. Mycotoxin production from fungi isolated from grapes. Letters in Applied Microbiology. 2001;**32**:240-242. DOI: 10.1046/j.1472-765X.2001.00897.x

[3] Tjamos SE, Antoniou PP, Kazantzidou A, Antonopoulos DF, Papageorgiou I, Tjamos EC. *Aspergillus Niger* and *aspergillus carbonarius* in Corinth raisin and wine-producing vineyards in Greece: Population composition, ochratoxin A production and chemical control. Journal of Phytopathology. 2004;**152**:250-255. DOI: 10.1111/j.1439-0434.2004.00838.x

[4] Iamanaka BT, Menezes HC, Vicente E, Leite RSF, Taniwaki MH. Aflatoxigenic fungi and aflatoxins occurrence in sultanas and dried figs commercialized in Brazil. Food Control. 2007;**18**:454-457. DOI: 10.1016/j.foodcont.2005.12.002

[5] Wei D, Wang Y, Jiang D, Feng X, Li J, Wang M. Survey of alternaria toxins and other mycotoxins in dried fruits in China. Toxins. 2017;**9**:200. DOI: 10.3390/toxins9070200

[6] Dachery B, Veras FF, Magro LD, Manfroi V, Welke JE. Exposure risk assessment to ochratoxin A through consumption of juice and wine considering the effect of steam extraction time and vinification stages. Food and Chemical Toxicology. 2017;**109**:237-244. DOI: 10.1016/j.fct.2017.09.013

[7] ElKhoury A, Rizk T, Lteif R, Azouri H, Delia M-L, Lebrihi A. Fungal contamination and aflatoxin B1 and Ochratoxin A in Lebanese wine–grapes and musts. Food and Chemical Toxicology. 2008;**46**:2244-2250. DOI: 10.1016/j.fct.2008.02.026

[8] Logrieco A, Ferracane R, Haidukowsky M, Cozzi G, Visconti A, Ritieni A. Fumonisin B2 production by *aspergillus Niger* from grapes and natural occurrence in must. Food Additives and Contaminants: Part A. 2009;**26**:1495-1500. DOI: 10.1080/02652030903148322

[9] Freire L, Braga PA, Furtado MM, Delafiori J, Dias-Audibert FL, Pereira GE, et al. From grape to wine: Fate of ochratoxin A during red, rose, and white winemaking process and the presence of ochratoxin derivatives in the final products. Food Control. 2020;**113**:107167. DOI: 10.1016/j.foodcont.2020.107167

[10] Mogensen JM, Larsen TO, Nielsen KF. Widespread occurrence of the mycotoxin fumonisin B2 in wine. Journal of Agricultural and Food Chemistry. 2010;**58**:4853-4857. DOI: 10.1021/jf904520t

[11] Guo W, Fan K, Nie D, Meng J, Huang Q, Yang J, et al. Development of a QuEChERS-based UHPLC-MS/MS method for simultaneous determination of six Alternaria toxins in grapes. Toxins. 2019;**11**(2):87. DOI: 10.3390/toxins11020087

[12] Ostry V, Malir F, Toman J, Grosse Y. Mycotoxins as human carcinogens—The IARC monographs classification. Mycotoxin Research. 2017;**33**:65-73. DOI: 10.1007/s12550-016-0265-7

[13] Serra R, Lourenço A, Alipio P, Venâncio A. Influence of the region of origin on the mycobiota of grapes with emphasis on *aspergillus* and *Penicillium* species. Mycological Research. 2006;**110**:971-978. DOI: 10.1016/j.mycres.2006.05.010

[14] Ubeda C, Hornedo-Ortega R, Cerezo AB, Garcia-Parrilla MC, Troncoso AM. Chemical hazards in grapes and wine, climate change and challenges to face. Food Chemistry. 2020;**314**:126222. DOI: 10.1016/j.foodchem.2020.126222

[15] Welke JE. Fungal and mycotoxin problems in grape juice and wine industries. Current Opinion in Food Science. 2019;**29**:7-13. DOI: 10.1016/j.cofs.2019.06.009

[16] Azaiez I, Font G, Mañes J, Fernández-Franzón M. Survey of mycotoxins in dates and dried fruits from Tunisian and Spanish markets. Food Control. 2015;**51**:340-346. DOI: 10.1016/j.foodcont.2014.11.033

[17] Terra MF, Prado G, Pereira GE, Ematné HJ, Batista LR. Detection of ochratoxin A in tropical wine and grape juice from Brazil. Journal of the Science of Food and Agriculture. 2013;**93**:890-894. DOI: 10.1002/jsfa.5817

[18] Poapolathep S, Tanhan P, Piasai O, Imsilp K, Hajslova J, Giorgi M, et al. Occurrence and health risk of Patulin and Pyrethroids in fruit juices consumed in Bangkok, Thailand. Journal of Food Protection. 2017;**80**:1415-1421. DOI: 10.4315/0362-028X.JFP-17-026

[19] Rahimi E, Jeiran MR. Patulin and its dietary intake by fruit juice consumption in Iran. Food Additives and Contaminants: Part B. 2015;**8**:40-43. DOI: 10.1080/19393210.2014.967814

[20] Ferranti LS, Fungaro MHP, Massi FP, da Silva JJ, Penha RES, Frisvad JC, et al. Diversity of aspergillus section Nigri on the surface of Vitis labrusca and its hybrid grapes. International Journal of Food Microbiology. 2018;**268**:53-60. DOI: 10.1016/j.ijfoodmicro.2017.12.027

[21] Lappa IK, Mparampouti S, Lanza B, Panagou EZ. Control of *aspergillus carbonarius* in grape berries by *lactobacillus plantarum*: A phenotypic and gene transcription study. International Journal of Food Microbiology. 2018;**275**:56-65. DOI: 10.1016/j.ijfoodmicro.2018.04.001

[22] Oliveri C, Bella P, Tessitori M, Catara V, Rosa RL. Grape and environmental mycoflora monitoring in old, traditionally cultivated vineyards on Mount Etna, southern Italy. Journal of the Science of Food and Agriculture. 2017;**97**:65-73. DOI: 10.1002/jsfa.7683

[23] Fanelli F, Cozzi G, Raiola A, Dini I, Mulè G, Logrieco AF, et al. Raisins and currants as conventional nutraceuticals in Italian market: Natural occurrence of Ochratoxin A. Journal of Food Science. 2017;**82**:2306-2312. DOI: 10.1111/1750-3841.13854

[24] Battilani P, Giorni P, Bertuzzi T, Formenti S, Pietri A. Black aspergilli and ochratoxin A in grapes in Italy. International Journal of Food Microbiology. 2006;**111**:S53-S60. DOI: 10.1016/j.ijfoodmicro.2006.03.006

[25] Battilani P, Magan N, Logrieco A. European research on ochratoxin A in grapes and wine. International Journal of Food Microbiology. 2006;**111**:S2-S4. DOI: 10.1016/j.ijfoodmicro.2006.02.007

[26] Kasfi K, Taheri P, Jafarpour B, Tarighi S. Characterization of antagonistic microorganisms against *aspergillus* spp. from grapevine

leaf and berry surfaces. Journal of Plant Pathology. 2018;**100**:179-190. DOI: 10.1007/s42161-018-0042-x

[27] Rousseaux S, Diguta CF, Radoï-Matei F, Alexandre H, Guilloux-Bénatier M. Non-botrytis grape-rotting fungi responsible for earthy and moldy off-flavors and mycotoxins. Food Microbiology. 2014;**38**:104-121. DOI: 10.1016/j.fm.2013.08.013

[28] Alsharif AMA, Choo Y-M, Tan G-H. Detection of five mycotoxins in different food matrices in the Malaysian market by using validated liquid chromatography electrospray ionization triple quadrupole mass spectrometry. Toxins. 2019;**11**:196. DOI: 10.3390/toxins11040196

[29] Mikušová P, Šrobárová A, Sulyok M, Santini A. Fusarium fungi and associated metabolites presence on grapes from Slovakia. Mycotoxin Research. 2013;**29**:97-102. DOI: 10.1007/s12550-013-0157-z

[30] Heperkan D, Gökmen E. Application of fourier transform infrared (FTIR) spectroscopy for rapid detection of fumonisin B2 in raisins. Journal of AOAC International. 2016;**99**:899-905. DOI: 10.5740/jaoacint.16-0156

[31] Abrunhosa L, Calado T, Venâncio A. Incidence of fumonisin B2 production by *aspergillus Niger* in Portuguese wine regions. Journal of Agricultural and Food Chemistry. 2011;**59**:7514-7518. DOI: 10.1021/jf202123q

[32] Bragulat MR, Abarca ML, Cabañes FJ. Low occurrence of patulin- and citrinin-producing species isolated from grapes. Letters in Applied Microbiology. 2008;**47**:286-289. DOI: 10.1111/j.1472-765X.2008.02422.x

[33] Santini A, Mikušová P, Sulyok M, Krska R, Labuda R, Šrobárová A. Penicillium strains isolated from Slovak grape berries taxonomy assessment by secondary metabolite profile. Mycotoxin Research. 2014;**30**:213-220. DOI: 10.1007/s12550-014-0205-3

[34] Ostry V, Malir F, Cumova M, Kyrova V, Toman J, Grosse Y, et al. Investigation of patulin and citrinin in grape must and wine from grapes naturally contaminated by strains of *Penicillium expansum*. Food and Chemical Toxicology. 2018;**118**:805-811. DOI: 10.1016/j.fct.2018.06.022

[35] Oztas E, Ozden H, Ozhan G. A preliminary survey of citrinin contamination in dried fruits, molasses and liquorice products in Turkey. Journal of Food and Nutrition Research. 2020;**59**:81-86

[36] Felšöciová S, Mašková Z, Kaˇ cániová M. Fungal diversity in the grapes-to-wines chain with emphasis on penicillium species. Journal of Food Science. 2018;**12**:379-386. DOI: 10.5219/882

[37] Alshannaq A, Yu J-H. Occurrence, toxicity, and analysis of major mycotoxins in food. International Journal of Environmental Research and Public Health. 2017;**14**:632. DOI: 10.3390/ijerph14060632

[38] Li M, Chen W, Zhang Z, Zhang Z, Peng B. Fermentative degradation of patulin by saccharomyces cerevisiae in aqueous solution. LWT- Food Science and Technology. 2018;**97**:427-433. DOI: 10.1016/j.lwt.2018.07.040 ·

[39] Susca A, Proctor RH, Mulè G, Stea G, Ritieni A, Logrieco A, et al. Correlation of mycotoxin fumonisin B2 production and presence of the fumonisin biosynthetic gene fum8 in *aspergillus nigerfrom* grape. Journal of Agricultural

and Food Chemistry. 2010;**58**:9266-9272. DOI: 10.1021/jf101591x

[40] IARC. Some naturally occurring substances, food items and constituents, heterocyclic aromatic amines and mycotoxins. In: Monographs on the Evaluation of Carcinogenic Risks to Humans. Lyon, France: World Health Organization; 1993. ISBN 978-92-832-1256-0

[41] Welke JE, Hoeltz M, Noll IB. Aspects related to the presence of toxigenic fungi in grapes and ochratoxin A in wines. Ciência Rural. 2009;**39**:2567-2575

[42] Zimmerli B, Dick R. Ochratoxin A in table wine and grape-juice: Occurrence and risk assessment. Food Additives & Contaminants. 1996;**13**:655-668. DOI: 10.1080/02652039609374451

[43] European Commission. Commission regulation (EC) No 195/2023 of 25 April 2023 on maximum levels for certain contaminants in food and repealing regulation (EC) No 1881/2006. Official Journal of the European Union L. 2023;**119**:103-157

[44] Dachery B, Manfroi V, Berleze KJ, Welke JE. Occurrence of ochratoxin A in grapes, juices and wines and risk assessment related to this mycotoxin exposure. Ciência Rural. 2016;**46**:176-183

[45] European Commission. Commission regulation (EU) No 2002/1370 of 5 august 2022 amending regulation (EC) No 1881/2006 as regards maximum levels of ochratoxin A in certain foodstuffs. Official Journal of the European Union. 2022;**L206**:11-14

[46] Amezqueta S, Schorr-Galindo S, Murillo-Arbizu M, Gonzalez-Peñas E, De Cerain AL, Guiraud J. OTA-producing fungi in foodstuffs: A review. Food Control. 2012;**26**:259-268. DOI: 10.1016/j.foodcont.2012.01.042

[47] Spadaro D, Lorè A, Garibaldi A, Gullino ML. Occurrence of ochratoxin A before bottling in DOC and DOCG wines produced in Piedmont (Northern Italy). Food Control. 2010;**21**:1294-1297. DOI: 10.1016/j.foodcont.2010.02.017

[48] Quintela S, Villarán MC, De Armentia IL, Elejalde E. Ochratoxin A removal from red wine by several oenological fining agents: Bentonite, egg albumin, allergen-free adsorbents, chitin and chitosan. Food Additives and Contaminants: Part A. 2012;**29**:1168-1174. DOI: 10.1080/19440049.2012.682166

[49] Labrinea EP, Natskoulis PI, Spiropoulos AE, Magan N, Tassou CC. A survey of ochratoxin A occurence in Greek wines. Food Additives and Contaminants. 2011;**4**:61-66. DOI: 10.1080/19393210.2010.539707

[50] Murillo-Arbizu M, Amézqueta S, González-Peñas E, Cerain D, López A. Occurrence of ochratoxin a in southern Spanish generous wines under the denomination of origin "Jerez-Xérès-Sherry and 'Manzanilla'Sanlúcar de Barrameda". Toxins. 2010;**2**:1054-1064. DOI: 10.3390/toxins2051054

[51] Quintela S, Villarán MC, de Armentia IL, Elejalde E. Occurrence of ochratoxin A in Rioja Alavesa wines. Food Chemistry. 2011;**126**:302-305. DOI: 10.1016/J.FOODCHEM.2010.09.096

[52] Fernandes PJ, Barros N, Câmara JS. A survey of the occurrence of ochratoxin A in Madeira wines based on a modified QuEChERS extraction procedure combined with liquid chromatography–triple quadrupole tandem mass spectrometry. Food

Research International. 2013;**54**:293-301. DOI: 10.1016/j.foodres.2013.07.020

[53] Leal T, Abrunhosa L, Domingues L, Venâncio A, Oliveira C. BSA-based sample clean-up columns for ochratoxin A determination in wine: Method development and validation. Food Chemistry. 2019;**300**:125204

[54] Altiokka G, Can N, Atkoşar Z, Aboul-Enein HY. Determination of ochratoxin A in Turkish wines. Journal of Food and Drug Analysis. 2009;**17**(6):467-473. DOI: 10.38212/2224-6614.2577

[55] Anli E, Çabuk B, Vural N, Pina EB. Ochratoxin A in Turkish wines. Journal of Food Biochemistry. 2005;**29**:611-623

[56] Var I, Kabak B. Occurrence of ochratoxin A in Turkish wines. Microchemical Journal. 2007;**86**:241-247

[57] Öncü Kaya EM. Development and validation of a SPE-UHPLC-fluorescence method for the analysis of ochratoxin-A in certain Turkish wines. Macedonian Journal of Chemistry and Chemical Engineering. 2019;**38**(2):161-170. DOI: 10.20450/mjcce.2019.1624

[58] Ponsone ML, Chiotta ML, Combina M, Torres A, Knass P, Dalcero A, et al. Natural occurrence of ochratoxin A in musts, wines and grape vine fruits from grapes harvested in Argentina. Toxins. 2010;**2**:1984-1996. DOI: 10.3390/toxins2081984

[59] Teixeira TR, Hoeltz M, Einloft TC, Dottori HA, Manfroi V, Noll IB. Determination of ochratoxin A in wine from the southern region of Brazil by thin layer chromatography with a charge-coupled detector. Food Additives and Contaminants: Part B. 2011;**4**:289-293. DOI: 10.1080/19393210.2011.638088

[60] Zhang X, Chen L, Li J, Zhu B, Ma L. Occurrence of ochratoxin A in Chinese wines: Influence of local meteorological parameters. European Food Research and Technology. 2013;**236**:277-283. DOI: 10.1007/s00217-012-1886-5

[61] Zhong QD, Li GH, Wang DB, Shao Y, Li JG, Xiong ZH, et al. Exposure assessment to ochratoxin A in Chinese wine. Journal of Agricultural and Food Chemistry. 2014;**62**:8908-8891. DOI: 10.1021/jf500713x

[62] Lasram S, Oueslati S, Chebil S, Mliki A, Ghorbel A. Occurrence of ochratoxin A in domestic beers and wines from Tunisia by immunoaffinity clean-up and liquid chromatography. Food Additives and Contaminants: Part B. 2013;**6**:1-5. DOI: 10.1080/19393210.2012.716453

[63] Di Stefano V, Pitonzo R, Avellone G, Di Fiore A, Monte L, Ogorka AZT. Determination of aflatoxins and Ochratoxins in Sicilian sweet wines by high-performance liquid chromatography with Fluorometric detection and Immunoaffinity cleanup. Food Analytical Methods. 2015;**8**(3):569-577. DOI: 10.1007/s12161-014-9934-3

[64] Stefanaki I, Foufa E, Tsatsou-Dritsa A, Dais P. Ochratoxin A concentrations in Greek domestic wines and dried vine fruits. Food Additives and Contaminants. 2003;**20**(1):74-83. DOI: 10.1080/0265203021000031537

[65] Burdaspal P, Legarda T. Occurrence of ochratoxin A in sweet wines produced in Spain and other countries. Food Additives and Contaminants. 2007;**24**:976-986

[66] Abrunhosa L, Paterson RR, Venâncio A. Biodegradation of ochratoxin A for food and feed decontamination. Toxins. 2010;**2**:1078-1099. DOI: 10.3390/toxins2051078

[67] Grazioli B, Fumi MD, Silva A. The role of processing on ochratoxin A content in Italian must and wine: A study on naturally contaminated grapes. International Journal of Food Microbiology. 2006;**111**:S93-S96

[68] Quintela S, Villaran MC, López De Armentia I, Elejalde E. Ochratoxin A removal in wine: A review. Food Control. 2013;**30**:439-445

[69] La Placa L, Tsitsigiannis D, Camardo Leggieri M, Battilani P. From grapes to wine: Impact of the vinification process on Ochratoxin A contamination. Food. 2023;**12**:260. DOI: 10.3390/foods12020260

[70] Lasram S, Mani A, Zaied C, Chebil S, Abid S, Bacha H, et al. Evolution of ochratoxin A content during red and rose vinification. Journal of the Science of Food and Agriculture. 2008;**88**:1696-1703. DOI: 10.1002/jsfa.3266

[71] Blesa J, Soriano JM, Molto JC, Manes J. Factors affecting the presence of ochratoxin A in wines. Critical Reviews in Food Science and Nutrition. 2006;**46**:473-478. DOI: 10.1080/10408390500215803

[72] Brera C, Debegnach F, Minardi V, Prantera E, Pannunzi E, Faleo S, et al. Ochratoxin A contamination in Italian wine samples and evaluation of the exposure in the Italian population. Journal of Agricultural and Food Chemistry. 2008;**56**:10611-10618

[73] Otteneder H, Majerus P. Occurrence of ochratoxin A (OTA) in wines: Influence of the type of wine and its geographical origin. Food Additives and Contaminants. 2000;**17**(9):793-798. DOI: 10.1080/026520300415345

[74] Gonçalves A, Palumbo R, Guimarães A, Gkrillas A, Dall'Asta C, Dorne J-L, et al. The route of mycotoxins in the grape food chain. American Journal of Enology and Viticulture. 2020;**71**:89-104. DOI: 10.5344/ajev.2019.19039

[75] García-Cela E, Crespo-Sempere A, Gil-Serna J, Porqueres A, Marin S. Fungal diversity, incidence and mycotoxin contamination in grapes from two agro-climatic Spanish regions with emphasis on aspergillus species. Journal of the Science of Food and Agriculture. 2015;**95**:1716-1729. DOI: 10.1002/jsfa.6876

[76] Melki Ben Fredj S, Chebil S, Lebrihi A, Lasram S, Ghorbel A, Mliki A. Occurrence of pathogenic fungal species in Tunisian vineyards. International Journal of Food Microbiology. 2007;**113**:245e250. DOI: 10.1016/j.ijfoodmicro.2006.07.022

[77] El Khoury A, Rizk T, Lteif R, Azouri H, Delia ML, Lebrihi A. Occurrence of ochratoxin A- and aflatoxin B1-producing fungi in Lebanese grapes and ochratoxin A content in musts and finished wines during 2004. Journal of Agricultural and Food Chemistry. 2006;**54**:8977-8982

[78] Fredj SMB, Chebil S, Mliki A. Isolation and characterization of ochratoxin A and aflatoxin B1 producing fungi infecting grapevines cultivated in Tunisia. African Journal of Microbiology Research. 2009;**3**:523-527

[79] Benaziz M, Hajji L, Jadouali S, Nait'Mbark H, Hajjaj H, Ainane A, et al. Mycotoxigenic fungi of wine grape in Meknes viniyards (Morocco). Pharmacology. 2021;**2**:566-575

[80] Chunmei J, Junling S, Qi'an H, Yanlin L. Occurrence of toxin-producing fungi in intact and rotten table and wine grapes and related influencing factors. Food Control. 2013;**31**:5-13

[81] Pérez-Ortega P, Gilbert-López B, García-Reyes JF, Ramos-Martos N, Molina-Díaz A. Generic sample treatment method for simultaneous determination of multiclass pesticides and mycotoxins in wines by liquid chromatography–mass spectrometry. Journal of Chromatography A. 2012;**1249**:32-40. DOI: 10.1016/j.chroma.2012.06.020

[82] Nistor AM, Cotan ŞD, Cotea VV, Niculaua AM. Analysis of aflatoxins in rustically wines from eastern Romania using the direct real time method (DART). BIO web of conferences. 2019;**12**:02026. DOI: 10.1051/bioconf/20191202026

[83] Paterson RRM, Lima N. How will climate change affect mycotoxins in food? Food Research International. 2010;**43**(7):1902-1914. DOI: 10.1016/j.foodres.2009.07.010

[84] Bezuidenhout SC, Gelderblom CP, Gorst-Allman RM, Horak W, Marasas FO, Spiteller G, et al. Structure elucidation of the fumonisins, mycotoxins from Fusarium moniliforme. Journal of the Chemical Society, Chemical Communications. 1988;**11**:743-745. DOI: 10.1039/C39880000743

[85] Gelderblom WCA, Jaskiewicz K, Marasas WFO, Thiel PG, Horak RM, Vleggaar R, et al. Fumonisins-novel mycotoxins with cancer-promoting activity produced by Fusarium moniliforme. Applied and Environmental Microbiology. 1988;**54**:1806-1811. DOI: 10.1128/aem.54.7.1806-1811.1988

[86] Hocking AD, Su-Lin LL, Benozir AK, Robert WE, Eileen SS. Fungi and mycotoxins in vineyards and grape products. International Journal of Food Microbiology. 2007;**119**(1-2):84-88. DOI: 10.1016/j.ijfoodmicro.2007.07.031

[87] Knudsen PB, Mogensen JM, Larsen TO, Nielsen KF. Occurrence of fumonisins B2 and B4 in retail raisins. Journal of Agricultural and Food Chemistry. 2010;**59**:772-776. DOI: 10.1021/jf103855x

[88] Qi TF, Renaud JB, McDowell T, Seifert KA, Yeung KKC, Sumarah MW. Diversity of mycotoxin-producing black aspergilli in Canadian vineyards. Journal of Agricultural and Food Chemistry. 2016;**64**:1583-1589. DOI: 10.1021/acs.jafc.5b05584

[89] Logrieco A, Ferracane R, Visconti A, Ritieni A. Natural occurrence of fumonisin B2 in red wine from Italy. Food Additives and Contaminants. 2010;**27**:1136-1141. DOI: 10.1080/19440041003716547

[90] Tamura M, Takahashi A, Uyama A, Mochizuki N. A method for multiple mycotoxin analysis in wines by solid phase extraction and multifunctional cartridge purification, and ultra-high-performance liquid chromatography coupled to tandem mass spectrometry. Toxins. 2012;**4**:476-486. DOI: 10.3390/toxins4060476

[91] Frisvad JC, Thrane U, Samson RA. Mycotoxin producers. In: Dijksterhuis J, Samson RA, editors. Food Mycology a Multifaceted Approach to Fungi and Food. Boca Raton: CRC Press; 2007. pp. 135-159

[92] Codex Alimentarius Commission. Codex Committee on Food Additives and Contaminants Maximum Level for Patulin in Apple Juice and Apple Juice Ingredients and Other Beverages. Codex Standard 235, 1. Rome, Geneva: FAO, WHO; 2003

[93] Broggi L, Reynoso C, Resnik S, Martinez F, Drunday V, Bernal ÁR. Occurrence of alternariol and alternariol

monomethyl ether in beverages from the Entre Rios Province market, Argentina. Mycotoxin Research. 2013;**29**(1):17-22. DOI: 10.1007/s12550-012-0147-6

[94] López P, Venema D, de Rijk T, de Kok A, Scholten JM, Mol HG, et al. Occurrence of Alternaria toxins in food products in the Netherlands. Food Control. 2016;**60**:196-204. DOI: 10.1016/j.foodcont.2015.07.032

[95] Zwickel T, Klaffke H, Richards K, Rychlik M. Development of a high performance liquid chromatography tandem mass spectrometry based analysis for the simultaneous quantification of various Alternaria toxins in wine, vegetable juices and fruit juices. Journal of Chromatography A. 2016;**1455**:74-85. DOI: 10.1016/j.chroma.2016.04.066

[96] Fernández-Cruz ML, Mansilla ML, Tadeo JL. Mycotoxins in fruits and their processed products: Analysis, occurrence and health implications. Journal of Advanced Research. 2010;**1**:113-122. DOI: 10.1016/j.jare.2010.03.002

[97] Hernandez MM, Menéndez CM. Influence of seasonality and management practices on diversity and composition of fungal communities in vineyard soils. Applied Soil Ecology. 2019;**135**:113-119. DOI: 10.1016/j.apsoil.2018.11.008

[98] Jiang C, Shi J, Zhu C. Fruit spoilage and ochratoxin a production by *aspergillus carbonarius* in the berries of different grape cultivars. Food Control. 2013;**30**:93-100. DOI: 10.1016/j.foodcont.2012.07.039

[99] Thiéry D, Louâpre P, Muneret L, Rusch A, Sentenac G, Vogelweith F, et al. Biological protection against grape berry moths. A review. Agronomy for Sustainable Development. 2018;**38**:15. DOI: 10.1007/s13593-018-0493-7

[100] Cozzi G, Pascale M, Perrone G, Visconti A, Logrieco A. Effect of Lobesia botrana damages on black aspergilli rot and ochratoxin A content in grapes. International Journal of Food Microbiology. 2006;**111**:S88-S92. DOI: 10.1016/j.ijfoodmicro.2006.03.012

[101] Chiotta ML, Ponsone ML, Torres AM, Combina M, Chulze SN. Influence of Planococcus ficus on aspergillus section Nigri and ochratoxin A incidence in vineyards from Argentina. Letters in Applied Microbiology. 2010;**51**:212-218. DOI: 10.1111/j.1472-765X.2010.02884.x

[102] Gil-Serna J, Vázquez C, González-Jaén MT, Patiño B. Wine contamination with ochratoxins: A review. Beverages. 2018;**4**(1):6

[103] Medina A, Medina A, Akbar A, Baazeem A, Rodriguez A, Magan N. Climate change, food security and mycotoxins: Do we know enough? Fungal Biology Reviews. 2017;**31**:143-154. DOI: 10.1016/j.fbr.2017.04.002

[104] Daou R, Joubrane K, Maroun RG, Khabbaz LR, Ismail A, El Khoury A. Mycotoxins: Factors influencing production and control strategies. AIMS Agriculture and Food. 2021;**6**(1):416-447. DOI: 10.3934/agrfood.2021025

[105] Perdoncini M, Sereia M, Scopel F, et al. Growth of fungal cells and the production of mycotoxins. In: Vikas B, Fasullo M, editor. Cell Growth. Vol. 3. Rijeka: IntechOpen; 2019. DOI: 10.5772/intechopen.86533

[106] Valencia-Quintana R, Milić M, Jakšić D, et al. Environmental research and public health review environment changes, aflatoxins, and health issues, a review. International Journal of Environmental Research and Public Health. 2020;**17**:7850

[107] Kizis D, Vichou AE, Natskoulis PI. Recent advances in mycotoxin analysis and detection of Mycotoxigenic fungi in grapes and derived products. Sustainability. 2021;**13**(5):2537. DOI: 10.3390/su13052537

[108] Shephard GS. Current status of mycotoxin analysis: A critical review. Journal of AOAC International. 2016;**99**(4):842-848

[109] Venâncio A. Detection and determination of Ochratoxin A in grape products. In: Barkai-Golan R, Paster N, editors. Mycotoxins in Fruits and Vegetables. San Diego: Academic Press; 2008. pp. 249-259

[110] Cao M, Wang J, Wang M, Yuan X, Zhang X, Ma J, et al. Simultaneous detection of 22 mycotoxins in grape by QuEChERS and ultra-high-performance liquid chromatography coupled with quadrupole/orbitrap high-resolution mass spectrometry. Journal of Future Foods. 2024;**4**(4):369-375. DOI: 10.1016/j.jfutfo.2023.11.009

[111] Arroyo-Manzanares N, Gamiz-Gracia L, Garcia-Campana AM. Determination of ochratoxin A in wines by capillary liquid chromatography with laser induced fluorescence detection using dispersive liquid-liquid microextraction. Food Chemistry. 2012;**135**(2):368-372. DOI: 10.1016/j.foodchem.2012.05.009

[112] Anfossi L, Giovannoli C, Baggiani C. Mycotoxin detection. Current Opinion in Biotechnology. 2016;**37**:120-126. DOI: 10.1016/j.copbio.2015.11.005

[113] Louppis AP, Constantinou MS. Application of a validated method for the identification and quantification of mycotoxins in wines using UPLC-MS/MS. Separations. 2022;**9**(4):102. DOI: 10.3390/separations9040102

[114] Kholová A, Lhotská I, Uhrová A, Špánik I, Machyňáková A, Solich P, et al. Determination of Ochratoxin A and Ochratoxin B in archived Tokaj wines (vintage 1959-2017) using on-line solid phase extraction coupled to liquid chromatography. Toxins. 2020;**12**(12):739. DOI: 10.3390/toxins12120739

[115] Dias JV, Nunes MGP, Pizzutti IR, Reichert B, Jung AA, Cardoso CD. Simultaneous determination of pesticides and mycotoxins in wine by direct injection and liquid chromatography-tandem mass spectrometry analysis. Food Chemistry. 2019;**293**:83-91

[116] Appell M, Evans KO, Jackson MA, Compton DL. Determination of ochratoxin A in grape juice and wine using nanosponge solid phase extraction clean-up and liquid chromatography with fluorescence detection. Journal of Liquid Chromatography & Related Technologies. 2018;**41**(15-16):949-954. DOI: 10.1080/10826076.2018.1544148

[117] Medina Á, Rodríguez A, Magan N. Climate change and mycotoxigenic fungi: Impacts on mycotoxin production. Current Opinion in Food Science. 2015;**5**:99-104. DOI: 10.1016/j.cofs.2015.11.002

[118] Rahman HU, Yue X, Yu Q, Xie H, Zhang W, Zhang Q, et al. Specific antigen-based and emerging detection technologies of mycotoxins. Journal of the Science of Food and Agriculture. 2019;**99**(11):4869-4877. DOI: 10.1002/jsfa.9686

[119] Lv Z, Chen A, Liu J, Guan Z, Zhou Y, Xu S, et al. A simple and sensitive approach for ochratoxin A detection using a label-free fluorescent aptasensor. PLoS One. 2014;**9**(1):e85968. DOI: 10.1371/journal.pone.0085968

[120] Agriopoulou S, Stamatelopoulou E, Varzakas T. Advances in analysis and

detection of major mycotoxins in foods. Food. 2020;**9**(4):518. DOI: 10.3390/foods9040518

[121] Visconti A, Perrone G, Cozzi G, Solfrizzo M. Managing Ochratoxin A risk in the grape-wine food chain. Food Additives & Contaminants Part A. 2008;**35825**(2):193-202. DOI: 10.1080/02652030701744546

[122] Abarca ML, Bragulat MR, Castellá G, Cabañes FJ. Impact of some environmental factors on growth and ochratoxin A production by aspergillus Niger and aspergillus welwitschiae. International Journal of Food Microbiology. 2019;**291**:10-16

[123] Gómez C, Bragulat MR, Abarca ML, Mínguez S, Cabañes FJ. Ochratoxin A producing fungi from grapes intended for liqueur wine production. Food Microbiology. 2006;**23**:541-545. DOI: 10.1016/j.fm.2005.09.007

[124] Varga J, Kozakiewicz Z. Ochratoxin A in grapes and grape-derived products. Trends in Food Science and Technology. 2006;**17**:72-81. DOI: 10.1016/j.tifs.2005.10.007

[125] Leong SL, Hocking AD, Scott ES. Aspergillus species producing ochratoxin A: Isolation from vineyard soils and infection of Semillon bunches in Australia. Journal of Applied Microbiology. 2007;**102**:124-133

[126] Abrunhosa L, Inês A, Rodrigues AI, Guimarães A, Pereira VL, Parpot P, et al. Biodegradation of ochratoxin A by Pediococcus parvulus isolated from Douro wines. International Journal of Food Microbiology. 2014;**188**:45-52. DOI: 10.1016/j.ijfoodmicro.2014.07.019

[127] Sun X, Niu Y, Ma T, Xu P, Huang W, Zhan J. Determination, content analysis and removal efficiency of fining agents on ochratoxin A in Chinese wines. Food Control. 2017;**73**:382-392

[128] Amézqueta S, Gonzalez-Penas E, Murillo-Arbizu M, de Cerain LA. Ochratoxin A decontamination: A review. Food Control. 2009;**20**:326-333

[129] Chen W, Li C, Zhang B, Zhou Z, Shen Y, Liao X, et al. Advances in biodetoxification of ochratoxin A-A review of the past five decades. Frontiers in Microbiology. 2018;**9**:1386

[130] Hathout AS, Aly SE. Biological detoxification of mycotoxins: A review. Annals of Microbiology. 2014;**64**:905-919

[131] Del Prete V, Rodriguez H, Carrascosa AV, De Las RB, Garcia-Moruno E, Munoz R. In vitro removal of ochratoxin A by wine lactic acid bacteria. Journal of Food Protection. 2007;**70**:2155-2160

[132] Piotrowska M. The adsorption of ochratoxin A by lactobacillus species. Toxins. 2014;**6**(9):2826-2839. DOI: 10.3390/toxins6092826

[133] Piotrowska M, Zakowska Z. The biodegradation of ochratoxin A in food products by lactic acid bacteria and baker's yeast. In Progress in biotechnology. In: Bielecki S, Tramper J, Polak J, editors. Food Biotechnology. Vol. 17. Amsterdam, The Netherlands: Elsevier; 2000. pp. 307-310

[134] Scott PM, Kanhere SR, Lawrence GA, Daley EF, Farber JM. Fermentation of wort containing added ochratoxin A and fumonisins B1 and B2. Food Additives Contaminants. 1995;**12**:31-40

[135] Bizaj R, Mavri E, Cus J, Raspor FA. Removal of ochratoxin A in Saccharomyces cerevisiae liquid cultures. South African Journal of Enology and Viticulture. 2009;**30**:151-155

[136] Csutorás C, Rácz L, Rácz K, Futo P, Forgó P, Kiss A. Monitoring of Ochratoxin A during the fermentation of different wines by applying high toxin concentrations. Microchemical Journal. 2013;**107**:182-184. DOI: 10.1016/j.microc.2012.07.001

[137] Petruzzi L, Sinigaglia M, Corbo MR, Campaniello D, Speranza B, Bevilacqua A. Decontamination of ochratoxin A by yeasts: Possible approaches and factors leading to toxin removal in wine. Applied Microbiology and Biotechnology. 2014;**98**(15):6555-6567

[138] Gambuti A, Strollo D, Genovese A, Ugliano M, Ritieni A, Moio L. Influence of enological practices on ochratoxin A concentration in wine. American Journal of Enology and Viticulture. 2005;**56**:155-162

[139] Solfrizzo M, Avantaggiato G, Panzarini G, Visconti A. Removal of ochratoxin A from contaminated red wines by repassage over grape pomaces. Journal of Agricultural and Food Chemistry. 2010;**58**:317-323

[140] Wang L, Liu X, Cai R, Ge Q, Zhao Z, Yue T, et al. Detoxification of Ochratoxin A by pulsed light in grape juice and evaluation of its degradation products and safety. Innovative Food Science and Emerging Technologies. 2022;**78**:103024

[141] Di Stefano V, Pitonzo R, Cicero N, D'Oca MC. Mycotoxin contamination of animal feedingstuff: Detoxification by gamma-irradiation and reduction of aflatoxins and ochratoxin A concentrations. Food Additives Contaminants. Part A Chemistry, analysis, control, exposure & risk Assess. 2014;**31**:2034-2039. DOI: 10.1080/19440049.2014.968882

[142] Di Gregorio MC, Neeff DV, Jager AV, Corassin CH, Carão ÁCP, Albuquerque R, et al. Mineral adsorbents for prevention of mycotoxins in animal feeds. Toxin Reviews. 2014;**33**:125-135. DOI: 10.3109/15569543.2014.905604

[143] Cosme F, Inês A, Silva D, Filipe-Ribeiro L, Abrunhosa L, Nunes FM. Elimination of ochratoxin A from white and red wines: Critical characteristics of activated carbons and impact on wine quality. LWT—Food Science and Technology. 2021;**140**:110838

[144] Filipe-Ribeiro L, Milheiro J, Matos CC, Cosme F, Nunes FM. Reduction of 4-ethylphenol and 4-ethylguaiacol in red wine by activated carbons with different physicochemical characteristics: Impact on wine quality. Food Chemistry. 2017;**229**:242-251. DOI: 10.1016/j.foodchem.2017.02.066

[145] Commission Regulation. Commission Regulation (EC) No 606/2009 of 10 July 2009 Laying Down Certain Detailed Rules for Implementing Council Regulation (EC) No 479/2008 as Regards the Categories of Grapevine Products, Oenological Practices and the Applicable Restrictions. Off. J. Eur. Union, 1-59, 606/20(No 479/2008). Brussels: Commission Regulation; 2009

[146] Karanfil T. Activated carbon adsorption in drinking water treatment. In: Activated Carbon Surfaces in Environmental Remediation. New York, New York, USA: Elsevier Ltd; 2006. pp. 345-373

[147] Galvano F, Pietri A, Bertuzzi T, Piva A, Chies L, Galvano M. Activated carbons: In vitro affinity for ochratoxin A and deoxynivalenol and relation of adsorption ability to physicochemical parameters. Journal of Food Protection. 1998;**61**:469-475

[148] Di Stefano V, Avellone G, Pitonzo R, Capocchiano VG, Mazza A, Cicero N, et al. Natural co-occurrence of ochratoxin A, ochratoxin B and aflatoxins in Sicilian red wines. Food Additives Contaminants. Part A. 2015;**32**:1343-1351

[149] Heshmati A, Ghadimi S, Ranjbar A, Mousavi KA. Changes in aflatoxins content during processing of pekmezas a traditional product of grape. LWT. 2019;**103**:178-185. DOI: 10.1016/j.lwt.2019.01.001

[150] Heshmati A, Mehri F, Nili-Ahmadabadi A, Khaneghah AM. Changes in aflatoxin content during the processing of vinegar obtained from grape. Avicenna Journal of Environmental Health Engineering. 2021;**8**(1):17-21

[151] Cosme F, Inês A, Ferreira B, Silva D, Filipe-Ribeiro L, Abrunhosa L, et al. Elimination of aflatoxins B1 and B2 in white and red wines by bentonite fining. Efficiency and impact on wine quality. Food. 2020;**9**:1789. DOI: 10.3390/foods9121789.doi:10.3390/foods9121789

Chapter 3

Grape Technology *vs.* Climate Change: A Success Story or a Nightmare?

Lucia Cintia Colibaba, Ionica Bosoi, Mărioara Pușcalău, Camelia Elena Luchian, Liliana Rotaru and Valeriu V. Cotea

Abstract

Climate change impacts significantly on the biology of horticultural species, including grapevines. As shown in many studies, temperature changes influence the spread of grapevine cultivation, the timing and progression of vegetation phenophases, and the overall quality of grape production. Long-term studies on the dynamics of vegetation phenophases in relation to environmental conditions provide insights into the quantification of climate change effects. By observing changes in development and duration of key phenophases, researchers assess shifts in grapevine growth patterns and adaptability to evolving climatic conditions. While higher temperatures may initially seem beneficial, the complex interactions between climate factors cannot fail to impact grapevine health and wine quality. For example, extreme heat or changed rainfall patterns all pose challenges to grape cultivation. This Romanian vineyard-based research indicated notable increases in the average annual temperature. With some annual values of over 2.5°C beyond the multiannual average, these trends suggest a growing favorability for quality wine production in the region—but is that really so? The temperature changes recorded and their influence on grapevine phenology and wine characteristics highlight the need for ongoing research and proactive measures to ensure long-term sustainability and resilience of grape cultivation in the face of climate change.

Keywords: climate change, grape technology, Romania, grape phenology, grape composition

1. Introduction

The unquestionable reality of climate change poses a challenge to traditions and practices in many areas of activity, the wine-growing world included. Given that wine is a particularly sensitive and supple agricultural product, there is a clear need for savvy producers to experiment with a range of adjustments that can withstand summer heatwaves, droughts, and milder winters, not to mention the ever more frequent harsh phenomena that are offshoots of climate change: forest fires, spring frosts, severe hailstorms, and flooding. There is compelling evidence that more disruptions are to be expected at an ever faster pace.

For close to three decades now, grape growers worldwide have been faced with significant changes in weather patterns, which, on the face of it, seems to have actually benefited regions never before associated with producing quality wines, such as England, that were quick to seize this unexpected opportunity of customizing local economies so as to join the wine world. By the same token, warmer growing seasons have facilitated the production of consistently exceptional wines in regions such as Champagne, Burgundy, Barolo, the Rhine, and Mosel valleys, where prime vintages used to be rare. Irrespective of such success, the character of these wines has unequivocally evolved, whether subtly or distinctly, due to the changing climate.

Studies carried out in the main vineyards in France have found an increase in the average temperature over the last 15 years of approximately 1°C, similar to both the minimum and maximum temperatures. These changes have already accelerated the course of the vegetative phenophases of the vine (differentiated according to the variety), causing an earlier ripening of the grapes of between 3 weeks and a month. Research done in Australia [1] indicated that grapes of different *Vinifera* cultivars zoned in Australian vineyards ripened, on average, 2 days earlier each year over the past 15 years. Also, other studies [2–4] highlight the gap in the development of the vegetative phenophases, including the ripening of the grapes that occur in the middle of the summer season, at the potential risk of altering the organoleptic characteristics of the grapes. Following studies carried out in Alsace, in Colmar, it was found that the period between destemming and harvesting moved forward and got shorter, while the ripening of the grapes took place under the conditions of increased temperature [5]. A study from INRA Bordeaux [6] believes that in view of the climate change trend, the zoning of varieties in wine plantations must take into account their thermal requirements in order to ensure the cycle of annual development: budding, flowering, veraison, ripening. In actual fact, the thermal conditions during the ripening period of the grapes are paramount for the grapes to reach the desired technological characteristics. Currently, the climate in the Bordeaux region is considered by the author to be intermediate, with an average air temperature (from April to October) of 16.5°C. The author believes that the average temperature increase of +2.4°C predicted for the end of the century remains within the limits of the thermal requirements of the main red varieties grown in the region. The edaphic factors, together with the climatic factors, influence the growth and fruiting phenophases of the vine, and implicitly the quality and quantity of the grape production.

Grapevine responds well in hilly regions in the Subcarpathian area of Romania and, to a lesser extent, in the plains. The altitude up to which the vine can be grown in Romania is 500–600 m, bearing in mind that with the increase in altitude, global radiation and the average air temperature decrease. The ideal conditions for vine cultivation are found at altitudes between 100 and 300 m.

In Romania, studies carried out by several authors [7–9] on the values of the limiting biotope factors for grapevine culture (as can be seen in **Table 1**) show that the level of climatic and eco-pedological factors is favorable for grapevine culture in well-defined cultivation areas in Romania, for both wine and table grape varieties. Climatic factors determine the area of vine cultivation and guide production directions.

Temperature represents the most important climatic factor and is expressed by the sum total of the temperature values, average and extreme, reached during vegetation and during the period of relative rest. The minimum thermal balances (global $\Sigma t°g$,

No.	Influencing factors		Measuring units	Restrictions	
				Minimal	Maximal
1.	Length of vegetation period		days	160	>200
2	Thermal balance	global	°C	2900	>4000-5000
		active		2700	>3800-4000
		useful		1000	>1500-1800
3.	Absolute minimal temperature for unprotected vine cultivation		°C	Table grapes	-18 ... -20
				Wine grapes	-20...-22
4.	Average temperature in warmest month		°C	18	>20–22
5.	Mean temperature at flowering		°C	15-17	30
6.	Mean temperature in soil		°C	5-6	25
7.	Real sunshine during vegetation		hours	1200-1300	>1800-2000
8.	Rainfall (annual average values)				
	Low trellising system		mm	450	800-900
	Semi-high and high trellising system		mm	500-550	800-900
9.	Relative humidity		%	50	80-90
10.	Climatic factors interaction	Real heliothermal index	-	1,3	>2
		Hydrothermal coefficient	-	0,6	3
		Bioclimatic index of vine	-	5	15
11.	Critical ecoclimatic factors	Hail	-	Absent or sometimes	
		Late and early hoar phenomena	-	Absent or at long intervals	
		Glazed frost	-	Absent or at long intervals	
		Excessive heat	-	Absent or at long intervals	

Table 1.
Eco-climatic conditions for the economic culture of grapevine as per [7].

active Σt°a, and useful Σt°u) for the economic culture of vine with extra-early and early ripening varieties are Σt°g = 2900°C, Σt°a = 2700°C and Σt°u = 1000°C. Medium ripening varieties require at least Σt°g = 3200°C, Σt°a = 2900°C, Σt°u = 1300°C, and those with late ripening: Σt°g = 3400°C, Σt°a = 3200°C, Σt°u = 1500°C. The average temperature of the hottest month (July) must be no less than 18°C for early varieties (table wine varieties), 20°C for average maturing varieties (quality wine varieties), and 22°C for late maturing ones (designation of origin wines). The minimum temperature during the flowering period (end of May, beginning of June, depending on variety and cultivation area) is 15°C for the opening of flowers and 17°C for pollen germination. The unprotected culture of the vine is possible only in areas where the absolute minimum temperatures in winter do not drop below −22°C in the case of wine varieties, or below −20°C for table grape varieties, provided that the frequency of such levels is a maximum of 1–2 years out of 10 [10].

Real insolation during the vegetation period must be at least 1200 hours for the culture of early varieties, 1400 hours for middle varieties, and 1500 hours for late varieties.

The annual precipitation must range from at least 450 mm for the low driving form to 500 mm for the semi-high forms, and 550 mm for the high driving forms, of which ⅔ needs to fall during the vegetation period. At precipitation values under 400 mm and 0.6 for the hydrothermal coefficient, grapevine culture is only possible under irrigation conditions. As a mesophytic plant, the humidity requirements for the grapevine vary during the vegetation period from one phenophase to another, subphases included. Thus, the phenophases of budding, and growing of shoots and berries are correlated with a higher value of the relative air humidity (70–80%), flowering with more than 55%, and the maturation of wood and berries with 50–60% humidity [10].

The areas designated for wine plantations must reach real heliothermic index (Ihr) values higher than 1.3, over 4 for the wine bioclimatic index (Ibcv), and over 3600°C for the oenoclimatic aptitude index (IAOe). Areas with recurrent climatic accidents need to be avoided: frosts in late spring and early autumn, hail, low temperatures during flowering, excessive heatwaves accompanied by drought during vegetation, or particularly cold currents during vegetative rest.

Research carried out worldwide on the impact of climate changes on the physiological and biochemical processes in the plant clearly points to repercussions for the development and duration of the main phenophases of the grapevine (budding, flowering, ripening, and grape ripening).

2. Case study: Romania, Vrancea County, Odobești vineyard

2.1 Multi-annual perspective

The climate of the area is temperate continental, with excessive nuances due to the presence of Eastern European air masses and Atlantic air masses from the west and northwest all year round, especially in the transition seasons, a climate similar to that of the Subcarpathian vineyards. The climatic elements in the southern part of Moldova generally tend not to show too great variations that would jeopardize the ecosystem for the culture of vines.

The Research and Development Station for Viticulture and Wine-making Odobești Meteorological Station located in the center of the vineyard, in the lower part of the Şarba plain (coordinates: 45°45″ north latitude, 27°03″ east longitude, and 182 m altitude), indicates a multi-year (1946–2019) average value of the annual temperature of 10.7°C, with +22.1°C for July and – 1.6°C for January. The average annual thermal amplitude of approximately 23.7°C corroborates with that of absolute extremes of over 63°C (between −23.7°C and + 39.4°C) to highlight the prominent degree of thermal continentalism.

Global solar radiation has high values, around 130 kcal/cm^2 (multiannual average), ranging between 120 kcal/cm^2 on northern exposures and 160 kcal/cm^2 on southern exposures. An increase of 20% favors the early initiation of some phenophases (budding, flowering), while a decrease of about 20–30% leads to a significant drop in the amounts of accumulated sugars, compounded by an increase in total acidity [11].

The real insolation in the Odobești vineyard has values between 1869.3 hours/year (1984) and 2578.6 hours/year (2012), of which 1509.9 hours during the vegetation period (multiannual value of the period 1970–2019), with maximum values of 1782.6 and 1745.9 in 2012 and 2018, respectively.

The global thermal balance for vineyards in Romania records values between 2700 and 3600°C, the active one between 2600 and 3500°C, and the useful one between 1000 and 1700°C [10]. In Odobești, the values (2974.2–3932.0°C) of the global heat balance from the vegetation period (1970–2019) indicate that the conditions for grape ripening are up to the VI and VII epochs. The active and useful thermal balance from the vegetation period of 2844.9–3749.3°C and 1188.8–1987.5°C, respectively, shows high annual variability, with a peak in 2012 (**Table 2**).

Year	Global thermal balance	Active thermal balance	Useful thermal balance	Year	Global thermal balance	Active thermal balance	Useful thermal balance
1970	3211.2	3086.7	1416.7	1996	3105.4	3096.8	1456.8
1971	3147.3	2978.7	1293.7	1997	3107.5	2984.1	1374.1
1972	3300.7	3198.7	1488.7	1998	3394.5	3394.5	1546.5
1973	3223.3	3145.6	1405.6	1999	3432.9	3370.5	1620.5
1974	3116.9	2966.2	1346.2	2000	3467.2	3461.3	1711.3
1975	3457.7	3374.7	1644.7	2001	3379.2	3407.4	1677.4
1976	2981.2	2928.4	1208.4	2002	3505.2	3424.2	1714.2
1977	3107.8	2942.1	1342.1	2003	3462.1	3373.9	1683.9
1978	2974.2	2848.8	1188.8	2004	3574.0	3142.1	1401.1
1979	3209.3	3081.3	1431.3	2005	3603.0	3165.5	1465.5
1980	2975.6	2844.9	1204.9	2006	3700.0	3191.5	1481.5
1981	3096.2	2995.6	1335.6	2007	3652.6	3579.4	1839.4
1982	3231.5	3066.1	1456.1	2008	3879.0	3473.2	1683.2
1983	3330.2	3259.7	1519.7	2009	3932.0	3502.2	1742.2
1984	3056.7	2892.8	1272.8	2010	3739.0	3432.4	1662.4
1985	3316.4	3251.0	1491.0	2011	3749.0	3314.3	1644.3
1986	3415.2	3281.4	1491.4	2012	3800.0	3737.5	1987.5
1987	3272.4	3110.0	1490.0	2013	3602.2	3523.1	1793.1
1988	3263.3	3151.5	1491.5	2014	3592.9	3508.1	1786.1
1989	3269.6	3246.5	1446.5	2015	3768.3	3688.9	1969.6
1990	3276.6	3210.8	1500.8	2016	3705.9	3688.0	1878.0
1991	3118.2	2970.6	1320.6	2017	3574.7	3478.9	1808.9
1992	3278.8	3125.7	1495.7	2018	3768.7	3749.3	1939.3
1993	3168.3	2902.9	1392.9	2019	3522.7	3377.8	1717.8
1994	3624.9	3582.8	1802.8	Multiannual mean	3395.5	3254.8	1552.3
1995	3331.8	3233.2	1553.2				

Table 2.
Thermal balance values (global, active, and useful) from the vegetation period in Odobești vineyard (1970–2019).

The average temperature of 20.9°C calculated for the first and second decades of June, that is, the period when the flowering phenophase occurs, proves favorable to the biological processes that condition the pollination and binding of the berries (**Table 3**).

The multiannual value of the average temperatures in July, August, and September (23.0°C, 22.9°C, and 17.4°C, to be precise) indicates a thermal climate favorable to the growth of the berries, veraison, and full maturity.

Year	Mean temp. of first 2 decades in June, °C	Mean temp. of warmest month (July), °C	Mean temp. of Aug., °C	Mean temp. of Sept., °C	Bioactive period (days)
1990	20.2	22.0	21.8	16.2	210
1991	19.2	21.9	20.2	16.5	192
1992	18.6	21.9	24.6	15.8	180
1993	20.1	20.7	20.9	16.2	174
1994	19.3	23.7	23.1	20.8	200
1995	21.1	24.0	21.6	15.7	187
1996	21.4	21.3	20.1	13.7	190
1997	20.3	21.2	19.8	14.3	200
1998	21.0	22.9	21.4	15.7	173
1999	22.2	23.6	21.2	18.0	183
2000	21.4	24.8	22.8	15.3	195
2001	19.6	24.3	24.2	16.9	192
2002	19.5	24.5	21.5	18.0	207
2003	22.3	22.1	23.1	15.9	199
2004	18.9	21.5	20.8	16.3	185
2005	17.9	21.5	21.2	17.6	187
2006	17.6	22.3	21.3	18.1	184
2007	23.1	25.4	23.4	17.4	205
2008	20.0	22.0	24.1	16.9	206
2009	22.0	23.7	22.9	18.9	217
2010	21.9	23.3	24.7	16.7	210
2011	21.5	22.6	22.1	20.0	216
2012	22.6	26.4	24.3	19.4	195
2013	21.1	23.1	23.5	16.4	214
2014	20.9	23.7	24.1	19.7	220
2015	22.9	25.0	24.5	20.2	205
2016	20.9	24.6	23.3	19.6	219
2017	21.3	22.6	23.9	18.9	220
2018	23.7	22.1	23.9	18.2	213
2019	23.1	22.1	23.9	19.0	218
Mean	20.9	23.0	22.9	17.4	199.8

Table 3.
Eco-climatic factors in Odobești vineyard between 1990 and 2019 as per [12, 13].

In recent years (2010–2019), one can observe an upward trend that helps obtain harvests with higher concentrations of sugars, flavors, and phenolic substances [10]. The duration of the bioactive period from 1990 to 2019 was, on average, 199.8 days.

As regards grapevine culture, the frequency of harmful absolute minimum temperatures under −20°C, which causes the soil to freeze up to 60–70 is of particular importance for unprotected vines. The analysis of the absolute minimum temperatures recorded in the winter months over a period of 50 years found the lowest temperature in the air to be −22.7°C on January 14, 1985, the very day when the absolute minimum on the ground was also recorded at −28.6°C. The average of the lowest absolute minimum temperatures in the air during the analyzed period was −15.2°C and at the ground surface - 19.1°C. These values confirm that the Odobești vineyard is located in a semi-protected vine cultivation area [12].

In the temperate continental climate, the vine requires an annual precipitation regime between 500 and 700 mm, of which at least 250–300 mm must be evenly distributed during the vegetation period as useful rainfall of over 10 mm [10]. In Odobești vineyard, the multiannual precipitation mean is 643.8 mm, of which 416.0 mm in the growing season, with a minimum of 164.9 mm in 1986 and a maximum of 693.4 mm in the year 2016 (**Table 4**).

Although there has not been a decrease in the amount of precipitation in recent years (2008–2019), with an annual average of 655.6 mm and 415.1 mm during the growing season, there appears to be an uneven distribution along the year, as well as an alternation between dry and rainy years.

The precipitation deficit appears especially in the winter period and the beginning of the vegetation period (April–May, with a direct impact on the initiation

Year	Rainfall, L/m^2		Year type	Year	Rainfall, L/m^2		Year type
	total	Veg. period			total	Veg. period	
1990	498.0	289.0	A little dry	2005	900.8	620.5	Excessively rainy
1991	804.0	634.8	Excessively rainy	2006	663.1	497.9	Very rainy
1992	637.9	395.9	Normal	2007	916.5	431.8	Excessively rainy
1993	610.4	331.7	Normal	2008	490.3	329.3	A little dry
1994	362.5	205.9	Excessively dry	2009	465.2	187.0	Excessively dry
1995	531.1	379.5	Normal	2010	775.0	500.2	Very rainy
1996	835.5	592.5	Excessively rainy	2011	509.4	409.2	normal
1997	782.1	512.0	Very rainy	2012	447.2	298.8	A little dry
1998	695.2	374.1	Normal	2013	813.4	555.4	Excessively rainy
1999	649.8	468.9	Rainy	2014	903.3	494.0	Very rainy
2000	451.1	287.9	Very dry	2015	650.6	330.0	Normal
2001	523.7	453.4	Rainy	2016	1049.0	693.4	Excessively rainy
2002	662.2	369.4	Normal	2017	655.2	407.6	Rainy
2003	409.7	264.3	Very dry	2018	561.8	405.8	Normal
2004	522.0	381.8	Normal	2019	546.8	370.8	Normal

Table 4.
Rainfall regime and the characteristics of the 1990–2019 period in the Odobești vineyard.

of the phenophases in the vines, as well as in July–August). The driest years were 1994 with 205.9 mm during the growing season and 2009, with only 187.0 mm, compared to the multiannual average (416.0 mm). Abundant precipitation was recorded in 2016, in the period preceding flowering (values almost triple compared to normal), which placed the harvest under manna attack, both on the leaves and on the bunches.

The analysis of precipitation over a period of 30 years (1990–2019) allows for a synthetic characterization of each year (**Table 4**). The data for the Odobeşti vineyard shows that there are excessively rainy and very rainy years (1–2 years) as well as excessively dry and very dry years (1–2 years) interspersed with normal years [12].

2.1.1 Synthetic ecoclimatic indices

To assess the heliothermal and water resources of a vineyard or wine-growing center, a series of synthetic ecological indicators are used that integrate the combined action of two or three climatic factors [9]: real heliothermal index—IHr; hydrothermal coefficient—CH [14]; bioclimatic index of the vine—Ibcv [15]; oenoclimatic aptitude index—IAOe [16]; De Martonne aridity index—IarDM [17]; Huglin index—IH [18]; night coolness index—IF [19].

In the Odobeşti vineyard, the average value of the real heliothermal index (IHr) from 1970 to 2019 is 2.4, with a minimum of 1.7 in 1976, 1978, 1980, and 1984, and a maximum of 3.5 in 2012, indicating an adequate temperature and light for the ripening of table and wine grape varieties (**Table 5**).

The CH has a multiannual average value of 1.3, with a minimum of 0.5 (1986 and 2009), which indicates extremely dry years, and a maximum of 2.1 (1978 and 1991), showing very rich years in precipitation.

The Ibcv with an average value of 7.2 is considered optimal for vine culture. In recent years (2000–2019) Ibcv recorded the most favorable values in 2000 (10.1), 2003 (11.0), 2009 (16.4), and 2012 (12.2), while the lowest value was recorded in 2005 (3.8).

IAOe with a multiannual average value of 4598 places Odobeşti vineyard in the area for white and red wines for current consumption, as well as quality wines with designation of origin. In certain years, this index places the vineyard among areas with a high degree of favorability for the production of quality white and red wines with designation of origin as well as table grape varieties.

The IarDM for said vineyard registered values between 16 (in 1994) and 49 (in 1972), and a multiannual average of 31, which indicates a semi-humid forest-steppe climate.

The IH, with a multi-year average of 2047, places the vineyard in the temperate climate class—IH3 (IH-1)*. However, in the last 10 years, the values went up to 2400 (2011–2020 average), thus placing the Odobeşti wine-growing area in the warm temperate climate class—IH4 (IH +1)*. The IF shows a multi-year value of 12.3, which qualifies as a wine-growing climate class with cold nights—IF_3 (IF +1) area*.

Overall, the values of the synthetic ecological indicators in the Odobeşti vineyard describe an area that is well-balanced and propitious for the cultivation of vines, with very good favorability for the cultivation of quality white wine varieties and good favorability for quality red wine varieties.

Year	IHr	CH	Ibcv	IAOe	Iar DM	IH	IF
1970	2.0	1.5	5.1	4296	36	1874	11.1
1971	1.9	1.8	4.5	4150	37	1801	10.5
1972	2.0	2.0	3.6	4123	49	1954	11.0
1973	2.1	0.8	10.1	4617	23	1905	11.8
1974	1.9	1.7	4.6	4123	33	1798	12.2
1975	2.2	1.0	7.5	4653	25	2139	13.7
1976	1.7	1.6	4.7	4093	39	1687	11.3
1977	2.0	1.3	6.4	4306	24	1650	11.1
1978	1.7	2.1	3.7	3947	42	1653	10.2
1979	2.1	1.7	4.9	4305	39	1877	12.4
1980	1.7	1.4	5.5	4067	38	1802	11.0
1981	2.0	1.5	5.4	4265	37	1768	12.2
1982	2.1	1.2	6.5	4398	27	1817	13.9
1983	2.2	1.2	6.7	4569	21	2051	12.7
1984	1.7	1.6	4.6	4013	38	1745	13.0
1985	2.3	1.0	8.4	4695	24	2038	11.2
1986	2.5	0.5	18.1	5033	17	2172	12.2
1987	2.1	1.3	5.8	4332	34	1985	12.9
1988	2.0	0.9	8.3	4464	31	1930	12.4
1989	2.0	1.4	5.4	4407	25	1949	12.6
1990	2.4	0.9	9.8	4781	23	2012	11.0
1991	1.8	2.1	3.4	3909	40	1748	12.0
1992	2.2	1.3	6.4	4454	31	1988	10.6
1993	2.1	1.1	7.0	4293	30	1864	11.2
1994	2.8	0.6	14.6	5168	16	2331	15.3
1995	2.4	1.2	7.2	4648	26	2090	11.4
1996	2.3	1.9	4.4	4311	41	1899	10.8
1997	2.1	1.7	4.8	4242	39	1767	9.8
1998	2.2	1.1	7	4674	33	2100	11.5
1999	2.6	1.4	6.2	4729	30	2089	14.0
2000	2.9	0.8	11.1	5117	21	2143	11.1
2001	2.6	1.3	6.4	4761	24	2118	12.8
2002	2.6	1.1	7.5	4791	30	2173	13.2
2003	2.7	0.8	11.0	4947	20	2158	11.5
2004	2.1	1.2	6.6	4487	25	1917	11.9
2005	2.0	1.9	3.8	4163	44	1882	13.4
2006	2.2	1.6	5.2	4423	32	1961	13.2
2007	3.2	1.2	7.8	5119	41	2346	12.2

Year	IHr	CH	Ibcv	IAOe	Iar DM	IH	IF
2008	2.5	0.9	8.5	4872	22	2187	11.9
2009	2.8	0.5	16.4	5166	21	2329	13.5
2010	2.3	1.5	5.2	4570	37	2190	12.1
2011	2.7	1.2	7.2	4789	24	2194	13.8
2012	3.5	0.8	12.2	5471	21	2597	12.7
2013	3.0	1.6	5.8	4877	37	2305	11.0
2014	2.8	1.4	6.1	4832	41	2329	13.8
2015	3.4	0.9	10.4	5318	28	2477	15.1
2016	3.2	1.9	4.9	4930	46	2451	14.0
2017	3.1	1.2	8.1	5048	29	2329	13.1
2018	3.4	1.1	8.8	5339	26	2551	13.1
2019	2.7	1.1	7.8	4824	24	2331	13.1
Average	2.4	1.3	7.2	4598	31	2051	12.3

Table 5.
Eco-climatics synthetic indices in the Odobești vineyard (1970–2019) as per [12, 13].

2.2 Recent climatic analysis 2020–2022

Maximum daily temperatures showed positive values yearly, with a steady increase from the winter to the summer season. The highest values of the maximum daily temperatures were recorded in the months of June and July. Daily minimum temperatures had negative values in December–February, with the lowest in January and the highest in July.

Average monthly temperatures showed an upward evolution in the first part of the year, with a maximum of 24.3°C in July (2021) and in August (2020 and 2022), followed by a downward trend, with a minimum in January.

Over the 2020–2022 period, the average air temperature was 12.3°C, 1.6°C higher than normal (10.7°C), and the average monthly temperatures were above the multiannual averages except for the month of April. A significant increase in average monthly temperatures was also observed in the winter period (December–February).

For the same time span, the maximum absolute temperature in the air reached 37.1°C (July 2021), and the absolute minimum (January 2021) showed −15.0°C in the air and – 17.8°C at the ground surface.

Based on the analysis of values for global heat balance (Σ °t global), active (Σ °t active), and useful (Σ °t useful) during the vegetation period, the following aspects emerge for the Odobești wine-growing area for 2020–2022:

- the global thermal balance, with values of 3344.0, 33586.9, and 3608.4°C over the 3 years of the study was favorable for the ripening of grapes;

- the active thermal balance with values of 3186.7, 3505.6, and 3543.5°C ensured the biological threshold for the growth and development of the vine;

- the useful thermal balance, ranging between 1566.7 and 1793.4°C, was beneficial for the ripening of grapes for both wine and table varieties.

Month	Global temperature °C				Active temperature °C			
	Multi-annual	2020	2021	2022	Multi-annual	2020	2021	2022
IV	346.2	367.1	262.6	347.4	254.6	302.2	105.3	266.1
V	529.4	489.6	510.5	549.6	513.2	489.6	510.5	549.6
VI	619.4	653.3	609.0	680.6	612.9	653.3	609.0	680.6
VII	695.8	729.1	753.5	742.5	689.3	729.1	753.5	742.5
VIII	682.3	751.8	708.6	752.2	674.9	751.8	708.6	752.2
IX	522.4	617.5	499.8	514.6	509.9	617.5	499.8	514.6
Σ	3395.5	3608.4	3344.0	3586.9	3254.8	3543.5	3186.7	3505.6

Table 6.
Sum of temperatures °C for vegetation period over 2020–2022 period.

The sums of global, active, and useful temperature degrees for the years in question exceeded the multiannual averages (**Tables 6** and 7).

Precipitation, along with temperature, is a very important climate factor for grapevine culture. The influence of humidity is of interest when one takes into account the ratio of precipitation level and air hygroscopicity.

According to several authors [7–8, 18, 20], vine culture is possible in temperate zone areas where the annual precipitation is between 400 and 700 mm, of which 250–300 mm is during the vegetation period.

Measurements in the Odobeşti vineyard show a multiannual average of annual precipitation of 643.8 mm, of which 416.0 mm during the vine vegetation period.

Over the 2020–2022 period, the amount of precipitation was both wanting and very unevenly distributed. The rainfall regime of 2020 was extremely deficient, with only 421.6 mm, of which 215.0 mm is in the vegetation period (**Table 8**), which makes it one of the driest years in the area under study.

Save June, when the amount of precipitation (81.6 mm) was close to normal (86.8 mm), most months saw amounts well below the normal values, prolonged drought settling in, and the precipitation deficit rising from 1 month to the next.

The amount of precipitation for 2021 was higher than in 2020 and 2022, but still below normal values, while the distribution along the vegetative rest period was disproportionate, with a deficit in February (−25.6 mm) and a slight surplus in January and March.

Month	Useful temperature °C			
	Multi-annual	2020	2021	2022
IV	71.0	82.1	15.3	76.1
V	218.4	179.6	200.5	239.6
VI	313.2	353.3	309.0	380.6
VII	375.6	419.1	443.5	432.5
VIII	361.7	441.8	398.6	442.2
IX	212.4	317.5	199.8	214.6
Σ	1552.3	1793.4	1566.7	1785.6

Table 7.
Sum of temperatures °C for vegetation period over 2020–2022 period (continuation).

Month	Rainfall (mm)				Hygroscopicity %			
	Multi-annual	2020	2021	2022	Multi-annual	2020	2021	2022
I	29.7	0.4	47.0	6.4	80.3	66.9	83.5	62.7
II	30.0	9.8	4.4	4.8	77.8	63.1	75.6	57.4
III	36.1	20.2	46.8	1.2	70.2	56.9	65.3	48.4
IV	50.8	5.2	41.0	65.8	65.8	39.0	65.3	58.1
V	80.3	55.6	22.8	31.6	65.9	60.1	63.8	54.5
VI	86.8	81.6	134.6	48.2	65.7	64.1	70.2	54.0
VII	83.0	28.2	40.0	51.0	65.5	57.0	63.2	53.5
VIII	64.4	13.0	45.0	31.6	64.9	50.1	64.3	59.4
IX	50.7	31.4	6.2	26.8	68.7	56.4	62.0	65.7
X	53.0	65.4	9.6	1.2	74.9	78.1	62.8	60.3
XI	42.3	12.2	12.0	50.8	79.3	82.0	75.2	83.4
XII	36.7	98.6	76.2	23.8	81.8	92.2	81.6	88.5
Annual	643.8	421.6	485.6	343.2	71.7	63.8	69.4	62.2
Veg. period	416.0	215.0	289.6	255.0	66.1	54.5	64.8	57.5

Table 8.
The rainfall regime and air hygroscopicity during the 2020–2022 period.

During the vegetation period, the amount of precipitation (289.6 mm) registered a deficit of 126.4 mm compared to the multi-year value for the Odobești vineyard (416.0 mm). The distribution of precipitation was uneven, with a significant deficit in May (−57.5 mm), moderate in July (−43.0 mm), and September (−41.1 mm), but also a substantial surplus in June (+47.8 mm). Specifically, the amount of precipitation recorded in 2021 was 485.6 mm below normal for the period (643.8 mm), which contributed to further extending the deficit from the previous year (2020). In turn, 2022 was extremely dry in the Odobești vineyard, with 343.2 mm of annual precipitation. The period of vegetative rest (January–March) lacked precipitation (−83.4 mm), while for the vegetation period (April–September) the deficit reached −161.0 mm compared to normal (416.0 mm), and was particularly significant in May (−48.7 mm). All this, on top of the already existing deficit (2021 and 2020 in particular), led to a sharp decrease in the water reserve in the soil.

Regarding the relative humidity of the air, it is known that vines normally carry out their growth and fruiting processes at values between 50 and 80%, those lower than 40% hinder photosynthesis, and below 20% block this activity.

During the analyzed period, the relative humidity of the air had values specific to the regions with a continental climate, which indicates the close correlation between this climatic element and the air temperature. Thus, in the years 2020 and 2022, the relative humidity of the air was 63.8% and 62.2%, respectively, being below the multiannual average.

2.2.1 Climatic indicators with a synthetic character

The combined action of the climatic factors is rendered *via* synthetic indicators (**Table 9**), which integrates the action of several ecoclimatic factors so as to facilitate

Climatic elements	Mean 1970-2019	2020	2021	2022	Mean 2020-2022
Global thermal balance, (Σt°g)	3395.5	3608.4	3344.0	3587.1	3513.2
Active thermal balance, Σt°a)	3254.8	3543.5	3186.7	3505.8	3412.0
Useful thermal balance, (Σt°u)	1552.3	1793.4	1566.7	1785.8	1715.3
Mean temp in first 2 decades of June	20.8	21.1	18.3	21.9	20.4
Mean temp in July, °C	22.2	23.5	24.3	24.0	23.9
Mean temp in August, °C	21.5	24.3	22.9	24.3	23.8
Mean temp in September, °C	17.1	20.6	16.7	17.2	18.2
Mean annual temp. t °C	10.7	13.0	11.3	12.7	12.3
Mean annual temp vegetation period, °C	18.2	19.7	18.3	19.6	19.2
Minimal temp in the air, °C	-22.7	-9.8	-15.0	-10.5	-11.8
Maximum temp in the air, °C	39.4	37.0	37.1	36.7	36.9
No of days with temp > 30 °C	28.9	67	46	55	56.0
∑ annual sunshine, (hours)	2194.9	2516.0	2309.5	2461.3	2428.9
Σ sunshine vegetation period, (hours)	1509.9	1725.0	1597.5	1679.5	1667.3
Σ annual rainfall, mm	643.8	421.6	485.6	343.2	416.8
Σ rainfall vegetation period, mm	416.0	215.0	289.6	255.0	253.2
Bioactive period, no. days	199.8	219	183	208	203.3
Real heliothermal index (Ihr)	2.4	3.1	2.5	2.9	2.8
Hydrothermal coefficient (CH)	1.3	0.6	0.9	0.7	0.7
Viticultural bioclimatic index (Ibcv)	7.2	15.5	9.6	12.4	12.5
Oenoclimatic suitability index (IAOe)	4598	5304	4745	5109	5053
Huglin heliothermal index (IH)	2047	2448	2123	2382	2318
Martone aridity index (Iar-DM)	31	18	23	15	18.7
Cooling nights index (IF)	12.3	14.6	11.0	11.4	12.3

Table 9.
Summary of the main climate components in the analyzed period compared with multiannual average.

the characterization of the climatic potential of a vineyard or wine-growing center, and also specify the requirements for the grape varieties studied.

The overall analysis of the main climatic factors in the Odobești vineyard over the period 2020–2022 compared to the multiannual averages highlights the following:

- an increase in the average annual temperature from 10.7°C (multiannual value) to 13.0°C (2020), and implicitly of the thermal balance values;

- an increase in average temperatures for July, August, and September;

- a reduction in annual and vegetation period precipitation amounts accompanied by very uneven distribution;
- twice the number of days with temperatures exceeding 30°C (56 days) compared to the multiannual value (28.9 days).

Systematic recording and processing of climate data for the period 2020–2022 provided the prerequisites for calculating binary and ternary climate indicators as well as multi-criteria climate indicators that describe the wine-growing climate of a wine-growing plot, wine-growing center, vineyard, or wine-growing region [21].

IHr values were between 2.5 and 3.1, exceeding the multiannual value and the numbers found in the literature (1.75–2.25), which shows an increase in heliothermal resources and optimal ripening conditions for late varieties in the southern part of Moldova. CH index was between 0.6 and 0.9, around the lower limit described in the literature (0.7–1.8), a testimony to insufficient humidity combined with increased temperatures. The Ibcv values for the same period were very high at 15.5 in 2020 and 12.4 in 2022, a fact that points to very high heliothermic resources and very low water resources and 9.6 in 2021, respectively. IAOE scored between 4745 (2021) and 5304 (2020), which is indicative of an area both opportune for quality white wines and boasting high favorability for red wines.

The DeMartonne aridity index saw values between 15 (2022) and 23 (2021), with an average of 18.7, which describes a semi-arid climate in the Odobești vineyard over the period in question.

The Huglin heliothermal index provides information related to the thermal potential in the culture of table and wine grape varieties with different grape ripening periods. For the Odobești vineyard, the sum total of the Huglin index values was 2448 in 2020, 2123 in 2021, and 2382 in 2022. As per these values, a warm climate class—IH5 was confirmed for the year 2020, with limits varying between 2400 and less than/equal to 3000, respectively, while a climate class IH4 specific to areas with warm temperate climate values above 2100 and less than/equal to 2400 was found to be the case for 2021 and 2022.

IF index is relevant for the ripening period of the grapes. It was obtained by summing the minimum temperatures of the respective month. The IF index was calculated for the month of September at 14.6 in 2020, 11.0 in 2021, and 11.4 in 2022. Such values fall within the IF_2 class range ($> 14 \leq 18$), corresponding to the climate class with temperate nights in 2020 when the ripening of medium varieties grapes is ensured. The IF_4 class range (< 12), corresponding to very cold nights in 2021 and 2022, when September is not favorable for grape ripening and production quality depends on solar radiation and insolation.

2.3 Grape cultivation in Odobești vineyard

The Research and Development Station for Viticulture and Oenology Odobeşti has been managing data on climate evolution since the 1940s and offers a perspective on how climate change affects local grape varieties and implicitly the quantity and quality of local wines. It also manages valuable databases on the phenology of the varieties, the ripening dates of the grapes of the different varieties in the area, the quality of the grapes and wines, etc. Based on these data, recommendations are made for choosing the right varieties to plant here and for applying optimal culture technologies.

The biological material that provides the corpus for this study is represented by five Romanian varieties of vines for white wines, some created at the Research and Development Station for Viticulture and Winemaking Odobeşti: Şarba, Băbeasca gri, Mioriţa and Vrancea, and Fetească regală, all Vitis vinifera.

2.3.1 ŞARBA grape variety

This was obtained by free fertilization of the Riesling Italian variety (Proles occidentalis). Authors: Popescu Gheorghe, Oşlobeanu Milu, Poenaru Ion, Bădiţescu Margareta. It was approved in 1972 [22]. The vegetation period is between 165 and 177 days. The budding season is late (end of April); the grape veraison begins at the beginning of August, and the full maturity of the grapes is achieved about 3–4 weeks after the Chasselas dore variety, in the second decade of September. The average production is 20 t/ha. At full maturity, the grapes accumulate a berry sugar content of 187 g/L, reaching up to 215 g/L in favorable years, and an acidity of 4.8–5.2 g/L sulfuric acid.

2.3.2 BĂBEASCĂ GRI grape variety

A variation of the Băbească neagră variety (Proles orientalis). Authors: Popescu Gheorghe, Oşlobeanu Milu, Poenaru Ion, Bădiţescu Margareta. It was approved in 1975 [22] Băbeasca gri variety has a vegetation period ranging between 190 and 205 days. Budding happens in the first half of April, the grape veraison takes place in the first half of August. It is a late-ripening variety, with grapes reaching full maturity at the beginning of October. It ensures high and constant productions that vary between 15 and 20 t/ha, and a good capacity to accumulate sugars. At full maturity, the grapes accumulate a content of sugars in the berries between 165 and 185 g/L, reaching almost 200 g/L in favorable years, and an acidity that varies between 6.0 and 6.7 g/L sulfuric acid.

2.3.3 MIORIŢA grape variety

This is a natural hybrid of the Coarnă albă variety (Proles orientalis). Authors: Gheorghe Popescu, Margareta Bădiţescu, Nicolae Varga, Zaharia Victoria. It was approved in 1980 [22]. The Mioriţa variety has a vegetation period between 167 and 182 days. Budding happens in the third decade of April, veraison is in the second decade of August, and full maturity of the grapes takes place in the last decade of September–early October. At full maturity, this late-ripening variety accumulates a sugar content of 165 g/L, reaching up to 180 g/L in favorable years, with high acidity values between 5.6 and 7.5 g/L sulfuric acid.

2.3.4 VRANCEA grape variety

Obtained by crossing the hybrid combination (Traminer x Armaş) with Fetească regală variety. Authors: Mihu Ghică, Bosoi Ionica, Puşcalău Marioara. It was approved in 2018 and patented in 2019. The vegetation period is between 164 and 175 days. Budding takes place in the second decade of April, blooming in the first decade of June, veraison in the first half of August, and full ripening in the second decade of September. The sugar accumulation potential in the must is 192–223 g/L, while total acidity is 5.48 g/L tartaric acid [23].

2.3.5 FETEASCĂ REGALĂ grape variety

This Romanian local grape variety is the result of natural hybridization between Fetească albă x Grasă de Cotnari [22] and, at the moment, is planted on the largest surface in Romania. Budding starts in the early days of April, veraison occurs in the first part of August, and full ripening is reached 3–4 weeks after the Chasselas doré variety. The grape production varies from one cultivated area to another, from 11 t/ha in Blaj to 27 t/ha in Odobeşti, with an average of 15–20 t/ha. It accumulates sugars from 170 to 180 g/L to 200–210 g/L, with a total acidity that must vary between 4.5 and 7.0 g/L sulfuric acid while the capacity for overripening is much reduced, with sugar accumulations not in excess of 220–235 g/L.

2.4 Climatic influences on grape phenophases and technological composition

Authors [24–27] were keen to study the behavior of grape varieties in their areas of origin under certain ecological conditions. The climatic changes in the last few years evidenced an average country-wide warming of 0.3°C, more pronounced in the eastern part, and a lower rainfall regime with uneven distribution all along the year.

The climatic conditions of 2020–2022 made possible in-depth research on the behavior of grape varieties created at S.C.D.V.V. Odobești (Şarba, Băbeasca gri, Miorița, and Vrancea), plus Fetească regală, the cumulative effect of stressful environmental factors included, as well as on their agrobiological and technological properties.

Research on the sequence and physiological completion of the phenophases in relation to the ecological factors of 2020–2022 highlights the fact that the vegetation phenophases were conditioned in many ways by the level and cumulative action of climatic factors and the hereditary specificity of each variety (**Table 10**).

In 2020, full ripening of the grapes occurred in the first and second decade of September, at average temperatures of 22.3°C, which led to the acceleration of physiological processes and faster ripening, too. In 2021, this phenophase occurred later, starting with the second decade of September for mid-ripening varieties (Şarba and Vrancea) and the third decade of September for late-ripening varieties (Băbeasca gri and Miorița), at average temperatures considerably lower than in the previous year (16.4°C).

In 2022, full ripening of the grapes was unusually early, *viz.* end of August—first decade of September, under the influence of high average temperatures of 22.6°C. During the 3 years of study, the first grapes to reach full maturity were Vrancea, the earliest on August 28 2022, and the latest on September 13 2021, to be followed by the Şarba variety 3–4 days later.

Miorița and Băbeasca gri varieties reached full maturity on September 9–112,022 (the earliest) and on September 23–252,021(the latest), ca 8–14 days after Fetească regală.

The duration of the grape ripening process was shorter, 32.0–39.5 days on average, with small differences among the varieties, which further confirms the influence of recent climatic changes in the Odobești wine-growing area (**Tables 11** and **12**).

As concerns harvest time, that is, the 1st decade of September in 2022 and the 2nd decade of September in 2020 and 2021 for the Vrancea and Şarba varieties, the 2nd decade of September in 2020 and 2022, and the 3rd decade of September in 2021 for late-ripening varieties (Băbeasca gri and Miorița), the analysis highlighted a real potential for the accumulation of sugars in the must.

Phenophases	Grape variety	2020		2021		2022	
		date	Useful Σ °t	date	Useful Σ °t	date	Useful Σ °t
Budding	Şarba	13.IV.	60.3	1.V	27.3	15.IV	54.5
	Băbească gri	12.IV.	56.4	28.IV	12.2	15.IV	54.5
	Mioriţa	14.IV.	60.3	4.V	50.3	16.IV	59.3
	Vrancea	13.IV.	60.3	2.V	38.3	16.IV	59.3
	Fetească regală	10.IV.	52.1	30.IV	17.7	15.IV	54.5
Flowering	Şarba	03.VI.	248.1	11.VI	282.6	31.V	273.8
	Băbească gri	04.VI.	259.9	14.VI	318.0	1.VI	284.0
	Mioriţa	05.VI.	268.5	14.VI	279.9	2.VI	291.6
	Vrancea	04.VI.	256.0	11.VI	271.6	29.V	253.1
	Fetească regală	03.VI.	256.3	10.VI	281.7	29.V	257.9
Veraison	Şarba	4.VIII	871.8	12.VIII	836.6	1.VIII	823.3
	Băbească gri	9.VIII	940.8	17.VIII	886.4	6.VIII	880.0
	Mioriţa	11.VIII	964.2	19.VIII	906.8	9.VIII	914.9
	Vrancea	3.VIII	850.5	10.VIII	809.9	27.VII	776.2
	Fetească regală	3.VIII	850.2	8.VIII	789.0	29.VII	803.2
Full maturity	Şarba	14.IX	528.3	17.IX	368.3	2.IX	454.0
	Băbească gri	17.IX	528.2	23.IX	320.9	11.IX	467.3
	Mioriţa	16.IX	568.5	25.IX	310.4	9.IX	389.1
	Vrancea	10.IX	483.9	15.IX	374.0	28.VIII	446.2
	Fetească regală)	9.IX	515.4	13.IX	383.2	26.VIII	404.6
Leaf fall	Şarba	16.XI	267.8	26. X	98.8	9.XI	252.3
	Băbească gri	16.XI	227.6	26. X	66.4	9.XI	181.6
	Mioriţa	16.XI	239.9	26. X	64.6	9.XI	203.0
	Vrancea	16.XI	321.7	26. X	120.3	9.XI	323.1
	Fetească regală	16.XI	334.6	26. X	141.2	9.XI	354.1
Length of vegetation period (days)	Şarba	216		178		207	
	Băbească gri	217		182		208	
	Mioriţa	215		179		208	
	Vrancea	216		176		207	
	Fetească regală	219		180		208	

Table 10.
The vegetation phenophases of studied varieties.

	Șarba				Băbească gri				Miorița			
	2020	**2021**	**2022**	**Mean**	**2020**	**2021**	**2022**	**Mean**	**2020**	**2021**	**2022**	**Mean**
Harvest date	16.09	20.09	10.09	-	7.09	23.09	14.09	-	17.09	24.09	15.09	-
Sugars, g/L	213	251	233	232	213	212	230	218	188	169	194	183
Total acidity g/L tartaric acid	5.5	7.8	5.2	6.2	7.	8.1	7.1	7.5	6.2	7.8	6.	6.7
pH	3.68	3.57	3.64	3.63	3.3	3.23	3.51	3.37	3.4	3.25	3.5	3.42
Gluco-acidimetric index	59	49	68	59	45	40	50	45	46	33	49	43
Total dry substance %	23.8	22.1	22.7	22.9	21.4	20.3	21.1	20.9	20.7	19.1	20.2	20
Total polyphenols g/L gallic acid	0.25	0.22	0.24	0.24	0.36	0.32	0.35	0.34	0.24	0.21	0.23	0.23
Total polyphenolic index (IPT)	2.8	2.4	2.7	2.6	4.0	3.4	3.8	3.7	3.5	2.7	2.6	2.9

Table 11.
Technological characteristics of studied varieties.

	Vrancea				Fetească regală			
	2020	2021	2022	Mean	2020	2021	2022	Mean
Harvest date	15.09	20.09	9.09	-	15.09	20.09	9.09	-
Sugars, g/L	205	203	221	210	192	183	197	191
Total acidity g/L tartaric acid	5.1	6.2	5.1	5.5	6.3	6.8	6.1	6.4
pH	3.62	3.52	3.6	3.61	3.5	3.4	3.5	3.53
Gluco-acidimetric index	61	50	66	59	47	41	49	46
Total dry substance %	23.1	21.3	22.6	22.3	22.1	20.6	21.5	21.4
Total polyphenols g/L gallic acid	0.32	0.27	0.30	0.30	0.29	0.23	0.27	0.,26
Total polyphenolic index (IPT)	3.9	3.5	3.6	3.7	3.5	2.8	3.0	3.1

Table 12.
Technological characteristics of studied varieties (continuation).

The sugar content of grapes is influenced by biological and ecological factors as well, as biological singularities of the variety, cultivation technology, insolation, temperature, humidity, soil characteristics, etc. As a result, the chemical composition of the must reveals different accumulations both from one variety to another and from 1 year to another. As seen in **Tables 11** and **12**, all grape varieties boast sugar concentration values higher than those quoted in the literature [28].

The acidity of the must in the four varieties stayed within normal limits in 2021 and scored much lower in 2020 and 2022 as a result of the lack of precipitation and the very high temperatures during harvest and ripening. Of the four varieties, the Băbeasca gri variety got the highest total acidity content, with an average value of 7.53 g/L tartaric acid, followed by Miorița, with an average of 6.73 g/L tartaric acid.

The balance between sugars and acidity of the must is also evidenced by the value of the gluco-acidimetric index. This index scored very high in 2020 and 2022 when the very low acidity of the must be varied between 46 for Băbeasca gri and 68 for Şarba. In 2021, deemed as climatically normal, the gluco-acidimetric index saw lower values, i.e., between 33 for Miorița and 50 for Vrancea, which is a guarantee for quality wines. The pH of the must had similar average values for the Băbeasca gri and Miorița varieties (3.37–3.42), scoring slightly lower in 2021 for Şarba (3.73), while Vrancea had values close to the control variety (3.53).

The total dry substance content (%) of the most was higher for Şarba, between 22.16–23.82%, averaging 22.92%, followed by Vrancea, with an average value of 22.39%, both above the Fetească regală variety (21.47%). The lowest values were recorded for Băbeasca gri (20.98%) and Miorița (20.03%). One should note that in the severe drought conditions of 2020 and 2022, the total dry matter content was higher, which signals a degree of berry dehydration, also highlighted by the lower moisture content (76.18–79.75%).

The total polyphenols, shown in grams/L of gallic acid, were higher for Băbească gri (0.32–0.36 g/L gallic acid) than the control variety, while Miorița totaled lower values (0.21–0.24 g/L gallic acid). The total polyphenolic indices (IPT) had the same distribution by varieties and harvest years, i.e., higher values for Băbeasca gri (3.4–4.0) and Vrancea (3.5–3.9), and lower values for the Miorița (2.6–3.5) and Şarba (2.4–2.8) varieties, which have green-yellowish skin and pulp.

3. Conclusions

The results compellingly show that early development stages already set in motion a general advancement of grapevine phenology and a curtailing of the growth period. Such an impact, corroborated by the further increase in the frequency and intensity of extreme climate events during sensitive phenological phases, is likely to have adverse effects on both the final yield and its quality and the suitability for grapevine cultivation of each region. Phenology advancement is expected to perturb the ripening period, as grape maturity occurs earlier during the hottest part of the vegetative cycle, customarily the warmest part of summer. The above is boosted by extremely high-temperature regimes insofar as they interfere with biochemical and physiological processes, more specifically berry sugar-acid and flavonoid levels, as well as color and aroma.

A trend shift between different white varieties as concerns the date of harvest was demonstrated and the data was pooled with sugar content increases and acidity reductions. We posit therefore that increased probability for unbalanced wines due to higher sugar and lower acid concentrations in the grape can be correlated with the predictable escalation of the number of early harvests (plausible for white varieties) and the corresponding reduction in the number of late harvests.

In conclusion, in Odobești vineyard an increase in average annual temperatures was recorded by 0.6–2.3°C, an uneven distribution of water resources, recording dry years, such as 2003 (−234 mm), 2009 (−178 mm), 2020 (−222 mm), and 2022 (−301 mm) against an average of 643 mm and rainy years such as 2005, 2007, 2014 but especially 2016 with an excess of over 400 mm.

The full ripeness of the grapes evolved according to the variety and was affected by the increase in temperatures and the increased frequency of days with maximum temperatures above 30°C recorded in July, August, and even in September.

All things considered, the wine industry is aware of the urgent need to make adjustments that take heed of climate change so as to preserve unique, centuries-old traditions and practices. By championing innovative approaches and sustainable practices, wine producers hope to continue creating exceptional wines while keeping the environment safe for generations to come.

Overall, the wine industry is recognizing the urgent need to adapt to climate change to preserve the unique traditions and practices that have been in place for centuries. By embracing innovative approaches and sustainable practices, wine producers hope to continue delivering exceptional wines while also safeguarding the environment for future generations.

Author details

Lucia Cintia Colibaba[1], Ionica Bosoi[2], Mărioara Pușcalău[2], Camelia Elena Luchian[1], Liliana Rotaru[1] and Valeriu V. Cotea[1]*

1 Iasi University of Life Sciences, Iasi, Romania

2 Odobești Vine and Wine Research and Development Station, Romania

*Address all correspondence to: vcotea@uaiasi.ro

References

[1] Webb LB, Whetton PH, Barlow EWR. Modelled impact of future climate change on the phenology of winegrapes in Australia. Australian Journal of Grape and Wine Research. 2007;**13**:165-175

[2] Lebon E. Changements climatiques: quelles conséquences prévisibles sur la viticulture. Actes des 6e Rencontres rhodaniennes, Institut rhodanien, Orange. 2002. pp. 31-36

[3] de García Cortázar-Atauri I. Adaptation du modèle STICS à la vigne (Vitis vinifera L.). Utilisation dans le cadre d'une étude du changement climatique à l'échelle de la France. SupAgro. [thesis]; 2006

[4] de García Cortázar-Atauri I, Brisson N, Gaudillere JP. Performance of several models for predicting budburst date of grapevine (Vitis vinifera L.). International Journal of Biometeorology. 2009;**53**(4):317-326

[5] Duchêne E, Schneider C. Grapevine and climatic changes: A glance at the situation in Alsace. Agronomy for Sustainable Development. 2004;**24**:93-99

[6] Bois B, Van Leeuwen C, Pieri P, Gaudillère JP, Saur E, Joly D, et al. Viticultural agroclimatic cartography and zoning at mesoscale level using terrain information, remotely sensed data and weather station measurements. Case study of Bordeaux winegrowing area. In: VIIth International Terroir Congress, Changins, Switzerland. Vol. 2. 2008. pp. 455-462

[7] Oprea Ş. Viticultura. Cluj Napoca: Ed. Academiei Pres; 2001

[8] Oşlobeanu M, Oprean M, Alexandrescu I, Magdalena G, Baniţă P, Jianu L. Viticultura generală şi specială. Bucureşti: Editura Didactică şi Pedagogică; 1980

[9] Ţârdea C, Dejeu L. Viticultură. Bucureşti: Editura Didactică şi Pedagogică; 1995

[10] Oşlobeanu M, Macici M, Magdalena G, Stoian V. Zonarea soiurilor de viţă-de vie în România. Bucureşti: Editura Ceres; 1991

[11] Constantinescu G, Oprea Ş. Le cépage, moyen principal de valorization des resources écologique. 1-er Simp. Constanţa: Int. sur l'Écologie de la Vigne; 1978. pp. 237-246

[12] Puşcalău M, Bosoi I, Dîrloman CA. Research on climate trends in the area of Odobeşti vineyard. Scientific Papers. Series B, Horticulture. 2021;**64**(1):334-341

[13] Ionica B, Marioara P. The phenological spectrum of some vine varieties cultivated in Odobeşti vineyard in the context of climate change. Lucrări ştiinţifice USV Iaşi, Seria Horticultură. 2021;**64**(1):171-176

[14] Seleaninov GT. Metodika selskohoteaistvenoi oţenki klima v subtropikah. Izd: Argrohidromet, Inst. R., Leningrad; 1936

[15] Constantinescu G, Donaud A, Dragomir E. Détermination de la valeur de l'indice bioclimatique de la vigne pour les principaux vignobles de RP Roumaine. Revue Roumaine de Biologie, série de Botanique. 1964;**9**:35-40

[16] Teodorescu Ş, Popa A, Sandu G. Oenoclimatul Romaniei. Bucureşti: Editura Ştiinţifică şi Enciclopedică; 1987

[17] De Martonne E. Une nouvelle function climatologique. L'indice d'aridité: La Meteorologie; 1926. pp. 449-458

[18] Huglin P. Nouveau mode d'évaluation des possibilités héliotermiques d'un milieu viticole, l-er Symp. Constanța: Int. sur l'Écologie de la vigne; 1978. pp. 89-98

[19] Tonietto J. Les macroclimats viticoles mondiaux et l'influence du mésoclimat sur la typicité de la Syrah et du Muscat de Hambourg dans le sud de la France: methodologie de carácterisation. Ecole Nationale Supérieure Agronomique, Montpellier, [thesis]; 1999

[20] Burzo I, Toma S, Olteanu I, Dejeu L, Delian L, Hoza D. Fiziologia plantelor de cultură. Vol. III. Bucureşti: Fiziologia viţei de vie, Editura Elisavaros; 2005. pp. 80-127

[21] Tonietto J, Carbonneau A. Le climat mondial de la viticulture et la liste des cépages associés. In: Système de Classification Climatique Multicritère (CCM) DES Régions á l'Échelle Géoviticole, Session du Grupe d'experts de l'OIV –Zonage Vitivinicole. Paris; 2000

[22] Rotaru L. Soiuri de viță de vie pentru struguri de vin. Iași Ed Ion Ionescu de la Brad; 2009

[23] Bosoi I, Rotaru L, Pușcalău M. Vrancea - new variety of vine for obtaining white wines of higher quality created at S.C.D.V.V. Odobeşti. Scientific Papers. Series B, Horticulture. 2020;**64**(2):133-137

[24] Bosoi I, Rotaru L, Colibaba LC, Pușcalău M. Agrobiological value of some white grape varieties created at Research and Development station for viticulture and Vinification Odobești Vrancea country, Romania. Journal of Applied Life Sciences and Environment. 2021;**54**(3(187)):289-297

[25] Colibaba LC, Cotea VV, Rotaru L, Tudose Sandu-Ville S. Studies of the influence of climatic changes on some grape varieties for white wines in Moldavian vineyards, Lucrări Ştiinţifice USAMV Iaşi, series B. Horticulture. 2013;**56**(1):275-278

[26] Pușcalău M, Bosoi I, Dîrloman CA. Preliminary results regarding the evaluation of the agrobiological and technological potential of some hybrid elite with biological resistance obtained at S.C.D.V.V. Odobeşti, Lucrări ştiinţifice USAMV Iaşi, series B. Horticulture. 2021;**64**(1):183-188

[27] Rotaru L, Colibaba LC. Phenological changes in some grape varieties for white wines in the north-east area of Moldova, Lucrări Ştiinţifice USV Iaşi, Seria. Horticultură. 2013;**56**(1):307-312

[28] Colibaba LC, Cotea VV, Rotaru L, Niculaua M, Nechita CB, Luchian EC. Studies on the compositional profile of wines obtained from Şarba grapes. Lucrări ştiintifice, Series B, Horticulture, USAMV Iaşi. 2014;**57**(1):161-164

Chapter 4

Opportunities and Challenges for Low-Alcohol Wine

Komal Sekhon and Qun Sun

Abstract

For centuries, diverse societies worldwide have fermented grapes and other raw materials to produce wine, elevating winemaking to an esteemed art form in numerous cultures and religions. Over time, globalization homogenized wine production, yielding a conventional approach known as traditional wine production. Recently, research on the adverse effects of alcohol consumption has spurred a trend toward low-alcohol wine, typically containing less than 8.5% alcohol by volume. This caters to health-conscious consumers and presents an economic opportunity for winemakers in an emerging market. Climate change has also impacted wine-grape cultivation, resulting in alterations of fruit quality and sugar accumulation. To address these challenges, viticulture practices are employed to mitigate photosynthetic product accumulation. Post-harvest, winemakers have various methods at their disposal to reduce alcohol content and maintain flavor balance. Producing low-alcohol wine poses challenges, including the risk of unwanted microbes and the need for vigilant monitoring during aging to prevent oxidation and spoilage. Despite the industry's millennia-old history, there remains ample room for innovation in low-alcohol winemaking. With shifting consumer preferences and climate change, the demand for lower-alcohol wines is poised to grow, necessitating ongoing research and innovative practices to create well-balanced wines that align with evolving consumer tastes and environmental conditions.

Keywords: non-traditional wine, climate change, terroir, health trends, excise taxes, sugar accumulation, ethanol extraction, alcohol removal, non-*Saccharomyces* yeast, fermentation kinetics, fermentation arrest, microbial spoilage

1. Introduction

With a long-recorded history, relics show humans have created and revered wine for thousands of years. Since the Neolithic period, humanity has intentionally created wine, with evidence of resinated wine products found in northern Iran dated to 5400–5000 BCE [1]. These wines were stored in resinated containers to take advantage of tree sap's preservation properties allowing for longer storage and less ullage. This practice continues today with retsina wine made in Greece [2, 3]. Wine can be traced back to many ancient civilizations including the Egyptians, Greeks, Romans, and Mesopotamians with notable social significance placed on the beverage. Around 3150 BCE, King Scorpion I of ancient Egypt prized wine for its status and ostentation and was buried with nearly 4500 liters filling three chambers from floor to ceiling

[4]. Throughout history, wine has constituted significant cultural and religious importance from Romans toasting for good health before voyages to modern-day Catholic communion with wine signifying the blood of Christ [5]. The mix between the art and science of winemaking developed alongside wine's cultural and religious prevalence.

Over time and alongside trade globalization, government-imposed taxes were levied on the sale and consumption of many alcoholic beverages and influenced winemaking practices. These tariffs are known as alcohol excise taxes and can vary based on the product classification. For example, in many European countries, wine products do not have any alcohol excise tax, although other alcoholic products even lower in alcohol content might [6]. There are also differences in taxation based on the type of product made, such as premium or sparkling wine having much higher taxes on average [7]. The alcohol content is also a significant factor in the wine trade as wines produced with lower alcohol by volume (ABV) fall into lower tax brackets and are thus more marketable to the consumer [8]. To remain within the ABV threshold, the wine may need to be adjusted to decrease the alcohol content with vineyard management, pre-fermentation, and/or post-fermentation practices. As such, with alcohol tax and maximizing revenue, many wines produced with similar methodology and alcohol content can be called "traditional" wine. In general, wine has become more alcoholic due to increased global temperatures and sugar accumulation, so more production interventions or lobbying efforts usually follow to maximize profitability while aligning with legal limitations [9]. There is an economic disincentive to producing wines with higher alcohol content.

Historically, the production of wine has changed over many years and regions. In the modern era, the speed of global transformation is unprecedented, and the industry must adapt to capture customer wants and to create production efficiencies. Different market niches and important terroir changes make the development of new wine practices a desired venture for the wine industry. A modern trend has arisen from research into the negative effects of alcohol consumption, with a more informed consumer base preferring wine with less alcohol. Low-alcohol wine, classified by the International Organization of Vine and Wine (OIV) as consisting of between 8.5% and 0.5% ABV, presents a new niche of consumers who are more health-conscious and/or hoping to gain benefits from wine consumption while minimizing alcohol consumption [10]. For beverages with less than 0.5% ABV, they can be labeled as non-alcoholic and are considered soft drinks. With these consumer preference changes comes an economic opportunity for winemakers to capitalize on the underdeveloped market of low-alcohol wine.

2. Modern-day trends and a health-conscious society

In more recent years, there has been a significant increase in consumer preference of toward low-alcohol wine, marking a notable trend in the wine industry. Consumer demand has changed as the population becomes more health-conscious and mindful towards consuming alcoholic beverages. More and more of the drinking age demographic are looking for alternatives to traditional wines with higher alcohol content [11]. While options of dealcoholized wine are available in the market, the taste profile of the wine may be very different from traditional wine and can deter regular drinkers from choosing this alternative [12]. Low-alcohol wine offers a lighter and more restrained drinking experience while maintaining the unique flavors and characteristics associated with wine [13]. This emerging trend reflects an increasing demand for alcoholic beverages aligned with wellness and moderation and opens a new niche for exploration from winemakers and wine drinkers alike in low-alcohol wine.

While wine is often associated with the negative health factors associated with alcohol consumption, low-alcohol wine may be beneficial in subtly decreasing the drawbacks by creating a healthier alternative. Multiple health institutes, such as the American Institute for Cancer Research (AICR) and the World Health Organization (WHO), correlate many health conditions with alcohol consumption [14]. The WHO considers decreasing alcohol drinking as a necessary public health initiative to combat the prevalence of alcohol-related cancers and the strain placed on the health institutes to combat these preventable diseases [15]. A study identified the causation effect alcohol consumption has on cardiovascular disease, stroke, and peripheral artery disease [16]. Current media and marketing campaigns emphasize alcohol beverages that have a lower cost for inebriation, this would be the opposite situation for low-alcohol wines [17]. Some negative aspects of traditional wine's alcohol content can be better managed if production and consumption of low-alcohol wine increases.

While alcohol is a major deterrent of consuming wine, the rich flavors and potential health benefits can highlight the positive effects of drinking a low-alcohol alternative. The French Paradox is a term used to describe the seemingly contradictory effects between the high-fat French diet and the relatively low incidence of cardiovascular disease among the French population [17]. Wine, particularly red wine, is viewed as one of the key factors contributing to this phenomenon. Red wine contains polyphenols, notably resveratrol, which possess antioxidant properties that can potentially reduce the risk of heart disease by protecting against oxidative stress and inflammation [18]. As a marketing strategy, wines that are lower in alcohol, thus the natural health benefits become more highlighted, can encourage a more health-conscious target audience to this functional food. Functional foods are scientifically demonstrated to have favorable effects on specific bodily functions, surpassing basic nutritional benefits, and contribute to an overall improvement in health, well-being, and disease risk reduction. A survey taken to determine if there is a market for functional wines (enhanced with resveratrol) showed the trend toward wellness products has an untapped potential in the wine industry from lower-alcohol wines [18].

2.1 "Healthy" choice: a double-edged sword

Low-alcohol wine has both an advantage and a disadvantage on the average consumer's perception of the product. While low-alcohol wine has the benefit of hopefully decreasing the alcohol consumption of the wine drinker, there have been studies that indicate the perceived "healthiness" of the product may encourage greater imbibing. This phenomenon can be attributed to a psychological aspect of societal perception and the desire for a more guilt-free drinking experience. When individuals believe a wine is lower in alcohol, they may consider it the healthier choice and therefore can indulge in greater quantities [19]. Similar to other "healthy" food alternatives, the perception creates a feeling of permission, as drinkers believe they can enjoy greater quantities of wine without the consequences of higher alcohol intake [20].

With product advertisement playing an influential role in profitability, how low-alcohol wine is labeled and marketed can be a greater risk for alcohol overconsumption if done improperly. With a broad range of alcohol contents that can fall into the category of "low-alcohol" wines, it is important to maintain the message that moderation is important for responsible drinking habits. The industry will need to cautiously market low-alcohol wine to gain a consumer base of health-cognizant drinkers, while not spreading the false message that low-alcohol wine can be consumed to no health detriment.

3. Climate change

The environment in which wine grapes grow is an important factor on the quality of wine that can be produced. The term "terroir" is used to describe the phenomenon of a flavor profile imbued onto the wine product from the soil, climate, vineyards' physical characteristics, and human management [21]. As the terroir of regions changes, so must the vineyard management practices and enological methods used to achieve optimal quality wine. The previously accepted measurements and classifications of winegrowing regions have become outdated with global temperature changes and water scarcity issues.

Climate change and recent global temperature increases have had a noticeable impact on grape quality and wine alcohol content. When contextualizing the issues faced by winegrowing regions, it becomes difficult to use water status to compare regions when water availability may differ from vineyard to vineyard. For this reason, temperature is the measurement more easily studied to determine which potential management practices to implement on a larger, regional scale. At temperatures above 35°C, the vines' photosynthetic pathways and gas exchange shut down to conserve water and energy when the plants are in prolonged high stress conditions [22]. In Carignane, Pinot Noir, Cabernet Sauvignon, and Tokay, temperatures from 32.5 to 40°C caused decreases in ovule fertility, berry weight, shoot growth, berry size, and berry cell count [23]. These high heat conditions worsen the quantity of berries and the volume of wine that can be produced. Another issue currently affecting excessively warm regions is the shutdown of the grapevines for long periods of time causing arrested phenolic and carbohydrate development. Besides the berries' external detriment in high heats, e.g. sun burn and shriveling, heat waves can decrease sugar accumulation of the berries so severe the grapes are not able to reach the desired dissolved sugar content (°Brix) [24]. There are few options available to the grape growers or wine producers when this occurs, and wine made from these grapes would be of poorer quality than desired. Lower sugar content would also produce lower-alcohol wine so these lower °Brix harvests may have to be mixed into another batch.

While excessive temperature increases cause decreases in vine metabolism and berry quality, moderate increases in temperature stimulate photosynthesis causing rapid sugar accumulation. This increase in sugar content translates to wine produced having higher alcohol content. Winemakers track the progress of the grape sugar content as a measurement of the juice °Brix to track the must weight and coordinate with the expected wine alcohol content [25]. When wine grapes are not properly tracked, overripe fruit tend to produce higher alcohol as the yeast has more sugar to digest, producing more ethanol [26].

A previous method of determining which grape varietals were suitable for regions was to calculate the growing degree days of the area and reference the Winkler Index [27]. The Winkler Index was a widely used scale with a baseline of 10°C developed by A. J. Winkler and M. A. Amerine to measure heat summation and correlate compatible grape cultivars with five classifications [28]. While the five categories (1 is little heat summation and 5 has the greatest) were relevant in the early 1900s, when the index was developed, current winegrowing regions have steadily gotten warmer and now surpass the upper limit of the Winkler Index. Previous compatible cultivars in set locations will have to shift toward areas with lower assigned Winkler Index values. For instance, regions where the Winkler Index previously classified certain varieties as too cold are now or will become warmer, thus the varieties could successfully be grown there [29]. The opposite also occurs where regions that had mild heat summation will no longer be able to sustain varieties not heat tolerant. A limitation of this scale is the narrow classification solely based on regions' temperatures. An ever-growing issue for wine-grape growers is water availability as natural water sources get diverted to support the growing human population and must be considered alongside climate. Temperature alone is not enough to classify which regions and grape varieties are compatible to achieve the greatest quality wine. The Winkler Index is an outdated scale to categorize winegrowing regions and should be updated to reflect the current and future conditions created by climate change.

The traditional method of assessing wine-grape suitability via the Winkler Index and the timeline of berry maturity previously followed has become less relevant with the impact of climate change. Winemakers and growers must work together to track sugar accumulation and develop management strategies to coordinate the best quality wine with optimal alcohol content.

4. Viticulture and enological practices for low-alcohol wine

4.1 Viticulture

The deliberate production of low-alcohol wine begins in the vineyards to decrease the fermentable sugars present in the grape must at harvest. Multiple management strategies can be utilized to reduce the total carbohydrates produced by the vines or to dilute the produced sugars per volume via increased yield.

An initial management strategy to decrease wine alcohol content is the selection of the grape variety and rootstock which maximizes fruit production. While certain cultivars are selected based on market preferences, other scion and rootstock varieties that have a lower sugar accumulation may be used for blending or specialty wines when developing low-alcohol alternatives [30]. These determinations must be made during vineyard planning with expectations of higher temperatures potentially increasing the popularity of these varieties. As climate change affects winegrowing regions and the compatible varieties, planting choices will change and evolve to take advantage of these new opportunities.

Certain vineyard management practices, such as increasing vigor or reducing photosynthetic ability of the vine, can also decrease the must weight. Maximizing vigor via enhancing bud load, lowering cluster thinning, and greater irrigation can help to increase the total yield and dilute the sugar content spread across the fruit [31]. Removal of the leaves, which produce the carbohydrates through their photosynthetic pathways, can also decrease the amount of fermentable sugars at harvest. Slowing down the carbohydrate concentration and delaying ripening can be done by canopy defoliation or trimming shoots [32]. In a field experiment testing mechanical defoliation, the leaf removal reduced ripening by 2 weeks and allowed greater phenolic development without the competition of sugar accumulation [33]. In the sensory evaluation conducted from this study's grapes, the wine was still considered well balanced and this may be from the decreased alcohol content being offset in flavor by the greater phenolics. Another method of reducing photosynthesis, while also being increasing drought tolerance, is the application of antitranspirants to the foliage and creating a thin, film barrier [34]. Vineyard management is the most important pre-harvest practice that can be implemented to produce lower-alcohol wine.

The final grape growing strategy used to decrease alcohol concentration is timing of the grape harvest. While it may seem that harvesting earlier would have a negative impact on the flavor profile due to reduced phenolic accumulation the differences are not enough to impact consumer preferences. In a sensory panel study, an early and late harvest was done on Cabernet Sauvignon, fermented separately, and the sensory panel found differences in the flavor profiles but no differences in likeability [35]. Multiple metrics are measured over time to track the ripening of the grapes and maximize quality, including the must titratable acidity (TA), pH, and °Brix. Early harvesting can also help blend wines of different phenolic and alcohol concentrations to produce lower-alcohol wines while maintaining a high-quality flavor profile. Additionally, harvesting earlier can reduce the strain on winery resources by fermenting smaller batches over a longer harvest period. Harvest timing can make a large impact on the wine's quality, quantity, and alcohol concentration.

Scion and rootstock selection is a decision that may affect the marketability of the vineyard for many years, so the decision to choose high vigor varieties that produce less alcoholic wines may see increased demand. The vineyard management strategies from bud break to final sugar accumulation play a large role in the preparation of the harvest for low-alcohol wine production. These practices must be skillfully executed to maintain the greatest wine flavor profile while reducing fermentable sugars. The last decision made in the vineyard is timing of the harvest, which is very influential on the must weight and subsequent wine production. These viticulture practices used individually or in combination can help to prepare the grapes for production of low-alcohol wine (**Table 1**).

4.2 Enology

4.2.1 Fermentation

After the harvest and preparation of the grape must, selection of enological practices can impact alcohol concentration in the final wine product. Two significant decisions that must be made regarding fermentation are the selection of yeast and manipulation of fermentation kinetics. Greater research into novel yeast options must be conducted to best suit the production of low-alcohol wine.

Method	Practice	Impact
Cultivar and vineyard planning	Scion/rootstock selection	Lower sugar accumulation Increased vigor Greater drought tolerance
Increasing vigor	Enhancing bud load Lower cluster thinning Increased irrigation	Increasing vigor Diluting present sugars
Reducing photosynthesis	Canopy defoliation Shoot trimming Foliar antitranspirants	Lower sugar accumulation Greater drought tolerance
Harvest timing	Measuring grape sugar and acidity	Maximizing flavor profile while avoiding excess sugar accumulation (TA, pH, °Brix)

Table 1.
Viticulture practices to produce low-alcohol wine.

Whereas centuries of yeast culture isolation and traditional winemaking practices have mostly selected for a single yeast species, the development of low-alcohol wine may not be conducive to simply changing the usage or the genetic potential of *Saccharomyces cerevisiae*. While a staple in any enologist's yeast selection, *S. cerevisiae* is not known for reduced alcohol production without necessary intervention to halt fermentation. The desire to continue using the same yeast species mainly stems from enologists' understanding of *S. cerevisiae* fermentation kinetics, while many other yeast species behave differently and are not as well documented [36]. While a viable option, potential modification of the *Saccharomyces cerevisiae* genome to decrease ethanol production can negatively affect wine quality and multiple modifications are needed to the genome to limit the negative byproducts that become more apparent as alcohol content decreases [37]. Although the usage of *S. cerevisiae* may still be suitable for low-alcohol wine.

Rather than manipulating the common yeast strains, there is potential value in branching out to other yeast species when developing low or non-alcoholic wines. When developing low-alcohol kiwi wine, using a mixture of *Saccharomyces* and *Wickerhamomyces anomalus* yeasts fermented to completion while also producing the positive flavor/odor profiles of the *Wickerhamomyces* yeast [38]. In a sensory panel survey of using a *non-Saccharomyces* yeast in production of non-alcoholic beer, while certain metabolites were quantitatively lower than the sensory threshold values, the trained sensory panelists were able to elicit the positive flavor profiles [39]. Similar to craft beer production, the development of low-alcohol wine can decrease the negative consequences of alcohol consumptions and allow the botanical components and polyphenols' flavor qualities to be better discerned by the consumer [40]. Although a deviation from the norm of winemaking, testing and development of more unconventional yeast strains for wine fermentation may have beneficial effects on the flavor profile when producing low-alcohol wines.

Another potential enological contribution to low-alcohol production is the purposeful arrest of yeast fermentation via insufficient nutrition. While multiple nutrition factors affect yeast digestion, the most significant measure used by enologists is the concentration of yeast assimilable nitrogen (YAN). Sources of YAN are mostly from ammonium and amino acids, and these are necessary to maintain fermentation kinetics to prevent "stuck" or incomplete fermentation [41]. While there is a portion of the necessary YAN present in the grape must, often supplemental nitrogen sources must be added to avoid starvation of the yeast [42]. While it is seen as an issue for traditional winemaking if fermentation becomes stuck due to lack of nutrition, for low-alcohol wine production, the yeast stopping alcohol production prematurely can help to reduce the human intervention or resources needed to reduce alcohol content [43]. Having a lower target YAN amount or simply refraining from adding supplemental nitrogen can naturally stagnate the yeast, although this method can be less predictable in terms of final alcohol or residual sugar content [44]. There may be lower alcohol content in the wine, but the balance of the flavor profile may also detract from the wine quality if too much residual sugar is present.

While insufficient nutrition may not be the goal for producing a well-balanced product, low-alcohol wine from arrested fermentation may also come from temperature shock or sterilization. If stopping fermentation with sufficient nutrition and fermentable sugars, temperature can be used as a control mechanism for manipulating fermentation kinetics. In a study of fermentation at varying temperatures, lower temperatures of 15–25°C saw greater efficiency at producing more ethanol per available sugar content and thus produced wines with greater total alcohol content

[45]. Although delayed in the yeast population growth initially, the lower temperature fermentations maintained greater total yeast concentration for longer and consumed the fermentable sugars much more rapidly compared to the 30 and 35°C fermentations [45]. Understanding the yeast population dynamics and tracking fermentation progression are valuable in determining the ideal conditions to stop fermentation. Once the residual sugars and alcohol content are at desired concentrations, rapid temperature changes can preserve these conditions. Cold shocking the yeast by bringing the wine below 10°C or pasteurizing with temperatures around 70°C will stop fermentation and kill the yeast [46]. Both methods require resources to manipulate the surrounding temperatures, which may not be viable for industrial production and rather chemical arrest of fermentation is a better technique. Although better suited for near-complete fermentations, the addition of sulfites and subsequent potassium sorbate will inhibit yeast growth and stabilize the wine [46]. Both the temperature and chemical manipulation of yeast populations to induce incomplete fermentation are viable methods to producing wine with lower alcohol content.

It has also been shown that extraction of the ethanol during fermentation may have beneficial effects on the fermentation kinetics. In a study comparing *de novo* aromatic compound synthesis and sensory effects of 2% ethanol removal by distillation under vacuum or carbon dioxide (CO_2) stripping, the results showed an increase from both methods in glycerol and isobutanol production with no sensory differences between the vacuum distillation extraction and wine produced without dealcoholization [47]. Although the vacuum distillation and CO_2 stripping processes required conditions that could increase yeast mortality (heat or high gas flow, respectively), the removal of the alcohol allowed for a greater production of aromatic compounds and indicate an increase in those pathways from the removal of medium-chain fatty acids [48]. The removal of ethanol during fermentation demands more resources and interventions but may create a much better flavor profile in the final wine product.

4.2.2 Post-fermentation

To produce low-alcohol wine, a prominent concern of winemaking is compromising between alcohol reduction while retaining the wine's sensory and flavor profile. Studies have shown flavor profile and aroma thresholds of alcoholic drinks differ when the alcohol is removed, thus a concern is prematurely arresting or reducing fermentation may produce undesirable flavors and it may be better to allow the fermentation to reach completion [39, 40]. If allowed to ferment completely or if fermentation produced greater alcohol than anticipated, additional manipulations can be made to remove excess alcohol accumulation.

As previously described, one viticulture strategy to reduce final alcohol content of the wine is early harvest. This limits the amount of sugar accumulated in the must and may have the added effect of retaining acidity, potentially contributing positively to the overall flavor profile. With decreased available sugar, fermenting an early harvest batch to completion would produce a wine with a lower alcohol content. The produced wine may not be received as well by the consumer for its reduced phenolics and increased acidity [39, 40]. A method winemakers can use to balance the wine is combining the early harvest wine batch with another from later in the season. The mixture of different batches can lower the final alcohol content and increase the overall wine quality. In areas experiencing greater warming and sugar accumulation or for smaller wineries, it can be a good strategy coordinated by the viticulture and enology sides to purposefully harvest an early batch to later blend with a batch harvested at the typical harvest period.

After fermentation, removing extra sugar and/or alcohol of the wine can help to produce a low-alcohol wine with better body, mouthfeel, and overall structure. Certain flavor and aroma compounds can only be produced during the fermentation process, so fermenting for longer may produce wine with a richer flavor profile. If fermentation is arrested, conducted with low-alcohol tolerant yeast, or not given sufficient nutrition, the residual sugar content could detract from the wine's flavor profile. Removal of the sugars or alcohol could be done via reverse osmosis, membrane filtration, or vacuum distillation [49]. The International Organization of Vine and Wine classifies the process of wine dealcoholizing into three separation techniques: partial vacuum distillation, membrane techniques, and distillation [50]. Partial vacuum evaporation decreases the air pressure in the container to less than the vapor pressure of ethanol and ethanol will then go into the gaseous state to be sucked out. OIV restricts these methods to only be used in wine absent of any organoleptic defects [50]. While some of these methods were previously unaffordable for many smaller-scale winery operations, recent warming patterns leading to increased sugar accumulation

Methods	Varieties	Impact
Use of Non-*Saccharomyces* Yeasts [51–53]	Chardonnay (*Metschnikowia pulcherrima* AWRI1149 followed by *S. cerevisiae*)	• Produced wines with a 0.9% (vol/vol) alcohol content; • Increased esters and higher alcohols; • Reduced volatile acids;
	Shiraz (*Metschnikowia pulcherrima* AWRI1149 followed by *S. cerevisiae*)	• Produced wines with a 1.6% (vol/vol) alcohol content; • Increased higher alcohols; • Reduced volatile acids;
	Trebbiano/Verdicchio (*H. uvarum Z sapae, Z bailii,* and *Z. bisporus)*	• Displayed a fermentation efficiency that was 62.5% (30.6% lower), while the *Zygosaccharomyces* species were all similar to *Saccharomyces*. • Increased residue sugar; • Increased ethyl acetate;
	Merlot (co-inoculated *M. pulcherrima* and *S. cerevisiae*)	• Decreased alcohol by 1%; • Increased ethyl acetate, total esters, total higher alcohols, and total sulfur compounds; • Received relatively high scores for red fruit and fruit flavor;
	Merlot (*S. uvarum*)	• Decreased alcohol by 1.7%; • Increased total higher alcohols; • Received relatively high scores for negative barnyard and meat flavor;
Modified *Saccharomyces* yeasts [54]	Tempranillo (glycolytically inefficient yeast, strain TP2A16)	• Decreased alcohol by 1%; • Increased glycerine; • Maintained volatile acid within the acceptable range;

Methods	Varieties	Impact
Nanofiltration (NF) [55]	Verdejo and Garnacha	• Dealcoholized alcohol by 2 degrees; • Observed no significant differences between the control and the filtered wines; • Identified the two-stage process without backflush as the most effective NF technique;
Pervaporation (PV) + nanofiltration (NF) [56]	Verdejo	• Dealcoholized alcohol by 1.7 degrees; • Exhibited an aroma content similar to that of the original must; • Demonstrated a taste similar to control wines; • The combination of PV and the two-stage NF yielded the best results;
Spinning cone column distillation [57]	13 red, 2 rose, and 4 white wines	• Reduced antioxidant activity due to loss of sulfur dioxide; • Showed elevated levels of phenolic compounds as a result of concentration;
Osmotic distillation (OD) [58]	Falanghina	• Dealcoholized at various alcohol content levels ranging from 9.8% to 0.3% ABV; • Preserved levels of total phenols, flavonoids, organic acids and total acidity; • Decreased volatile compounds with the progressive removal of alcohol;
Osmotic distillation (OD)-reverse osmosis (RO) [59]	Montepulciano d'Abruzzo	• Dealcoholized alcohol content from 8% to 5% ABV; • Showed a better retention of the main chemical properties and volatile compounds;
Membrane contactor (MC); Distillation under Vacuum (D) [60]	Rosé wine, Pelaverga Barbera	• Dealcoholized alcohol at 5% ABV; • Retained organic acids, cations, polyphenols, and anthocyanins; • Showed a reduction in volatile compounds;
Reverse osmosis-evaporative Perstraction (RO-EP) [61, 62]	Three wines sourced from industry	• Dealcoholized alcohol content from 5% to 0.5% ABV; • Enhanced color intensity, phenolics and organic acids; • Reduced certain fermentation volatiles, particularly ethyl esters;
	Cabernet Sauvignon	• Lower the alcohol content by between 1.8% and 2.5% ABV; • Exhibited reduced acidity, sweetness, bitterness, saltiness, and/or body; • Enhanced astringency;

Table 2.
Winemaking practices for low-alcohol wine production.

have persuaded many winemakers to invest in the necessary equipment. This equipment can easily be used both for the reduction of alcohol in their traditional wine and for the creation of a new platform of low-alcohol wine selections (**Table 2**).

5. The economics of production

The economic value to the consumers and producers is nuanced, with attractants and deterrents from the production and consumption of low-alcohol wine. The public perception, influenced by health organizations and general trends in alcohol consumption, has shown that there may be demand for wine products lower in alcohol and winemakers could capitalize on this opportunity. Conversely, the pricing of these more resource intensive products may not fit the producer and consumers' wishes and the profitability may be lower than traditional wine.

While consumers can gain health benefits from low-alcohol wine, wine producers can also reap the economic benefits. As the health risks of alcohol consumption strain public health infrastructure, many countries and organizations have proposed an increase in the excise tax rates of alcoholic beverages. In 2021, the World Health Organization (WHO) began considering the implications of doubling the excise tax on alcohol in an effort to limit the prevalence of alcohol-attributed cancers [15]. This hypothetical situation showed a decrease in other alcohol commodities but not wine due to many European Union (EU) WHO member states having no excise tax rate on wine. After the WHO study was published, a criticism emerged that alcohol consumption via wine is largely prevalent in these European countries with no excise tax and wine can make up one-third of the per capita alcohol consumption in these populations [6]. If there was an initiative to increase the excise tax rate on alcohol to decrease preventable, alcohol-attributable diseases, it stands to reason the countries with no or minimal wine excise tax would likely have to enact a wine excise tax. In Australia, a survey was conducted to gain an understanding of the drinking population and whether there is a market potential for low-alcohol wine if the excise tax rate increased [8]. The results of the survey found that there is market potential in consumers of wine and light beer but consumers of full-strength beer, spirits, and cocktails are less likely to purchase low-alcohol products. With current wine consumers more open to low-alcohol alternatives, the market potential for low-alcohol wine, and avoidance of increased excise tax on wine, producers should consider increasing their investment in the development of viable low-alcohol wine alternatives.

Although there may be market potential in capturing the niche of consumers who would prefer a low-alcohol alternative, many consumers would probably not change to drinking lower-alcohol wine without an economic incentive. With not enough market research in the existing low-alcohol wine options, these surveys regard other healthy food alternatives that are widely known. In a survey conducted on fast-food consumers and their perceptions on healthy food options, there was a clear trend that most people surveyed would prefer healthier menus and more sustainable items but only 7% of the participants' meals included healthier food choices already available [63]. The researchers contributed this lack of throughput with people's affinity for taste, price, and habit, which is also likely with consumers who claim to want more low-alcohol options.

Public perception of products which have had something removed may also influence consumer purchasing and drinking trends for low-alcohol wines. In another beverage product, milk, a survey found that average shoppers believed lower fat milk to be more expensive and whole milk to be a better source of calcium, which could be deterrents in their purchase of low-fat milk [64]. There does seem to be changes in the public's perception of these available alternatives as health trends highlight the positive effects of changing consumption habits. An example of this trend is the recent increase in coffee consumers choosing decaffeinated over traditional. Although the perception of decaffeinated coffee is similar to low-fat milk in its greater expense, the negative effects of traditional coffee's high caffeine amount deter consumers and they shift toward the healthier option with the premium price [65].

While there does seem to be a lack of representation of low-alcohol wine and the recommendations of health organizations would do well to encourage the production, customers would perhaps not pay for a premium wine price with lesser alcohol. The marketing of low-alcohol wine must also not oversell the health benefits and dupe the consumers that the products should be consumed with less regard for the alcohol content.

6. Challenges

While low-alcohol wine production may be a new niche that the industry can explore, there are considerable challenges that must be acknowledged. For both winemakers and winegrowers, the different methods that must be enacted to maximize wine quality while minimizing alcohol will likely differ from common practices. There are concerns over the viticulture practices in light of climate change and how to manage accordingly, and most of the expected challenges are with the winemaking side. Fermentation kinetics, microbial spoilage, instability, and aging all have problems that can arise from lower alcohol conditions.

As a support from the winegrowing side, it is best practice to plan whether a harvest will be used for low-alcohol wine prior to the grape growing season to better prepare the must for production. With a lower alcohol target, the desired sugar content of the grape must may be lower and thus the strategies explained above would be very beneficial to implement. As well, the post-fermentation practice of batch blending to reduce the final alcohol content can only be done if there was an early harvest batch to mix [66]. This early harvest would have higher acidity and lower sugar than growers would be expecting, so communication from the winemaking to the winegrowing teams is critical.

While producing the final product, the wine must be monitored to determine whether there are unwanted fermentation kinetics or microbial populations spoiling the wine. When deciding on the usage of non-*Saccharomyces cerevisiae* yeasts, most other wine yeasts are not as vigorous or competitive potentially becoming outcompeted by wild yeast [67]. Although there is a chance that the native yeast and microbes may create new, interesting flavors, there is a much larger probability that these wild yeasts will produce off-flavors, lower quality, or unusable wine. As mentioned before, the fermentation kinetics of *Saccharomyces* is well-known and well understood, while other wine yeasts less so. Wine yeasts that produce less alcohol or are less alcohol tolerant may arrest fermentation too early and leave more residual sugars than desired. An issue with too much sugar or not enough alcohol is the increased likelihood of microbial spoilage and instability. Alcohol is a natural sanitizer and can normally keep the wine microbe spoilage controlled, but may not in a situation of low alcohol with high sugars for the microbes to feed on [68]. In this case, *Saccharomyces cerevisiae* may still be used sequentially to digest the sugars further or other technologies to inactivate the microbes while also trying to retain the low-alcohol content. After fermentation, there is still a higher risk of microbial spoilage in the production and aging process. Meticulous attention to hygiene and sanitization is necessary to prepare the wine for aging, especially for long-term aging [69].

For the winemaking industry, the production of low-alcohol wine presents a new opportunity for market growth but it is not without limitations. For grape growing practices, low-alcohol wine has some methods that can better prepare the harvest for the fermentation process. While the downsides and the need for decreasing alcohol consumption are a social issue, a wine industry issue is that alcohol prevents many issues during fermentation that must then be carefully monitored to create a high-quality product. Low-alcohol wine production should be undertaken after determining the risks and having a strategy to maximize the quality and economic gain while minimizing any product loss.

7. Future work

While there is much research done into possible avenues for producing low-alcohol wine from traditional wine methods, there is still much to be explored to discover better practices and create efficiencies. While the viticulture practices are well known to lower sugar accumulation, mostly discovered from general practices needed to grow in warmer climates, there is insufficient research into enological practices in low-alcohol wine production.

As explained previously, many enological practices for traditional wine do not correlate well with low-alcohol production and studies into yeasts, fermentation kinetics, and industry-scale production are needed. *Saccharomyces cerevisiae* has been extensively researched and its fermentation kinetics in all wine varieties has been documented, which makes it a staple for any enologist. When branching into alternative yeast species, there is much less known about their respective fermentation kinetics. More depth of knowledge for these yeasts would not only help with low-alcohol wine production, but also with expanding the repertoire of winemakers to create new flavor profiles [70]. While it may be more comfortable for winemakers to continue using *Saccharomyces cerevisiae* for fermentation and then undergoing either batch mixing or ethanol extraction, this method is both resource/time intensive as well as unknown whether the flavor profile would be adequately balanced [39, 67]. Traditional winemaking has determined certain aroma or odor thresholds for both positive and negative flavor compounds, yet the removal of alcohol can lower this threshold. This can either be beneficial by highlighting more desired flavors, or detrimental with increasing the spoilage flavors or changing the flavor profile too far from consumers' expectations of certain varieties or wine styles.

In combination with the enological research necessary for better low-alcohol wine production, there also is a need for more studies using large-scale production. Many of the studies conducted and published on reducing alcohol content are scaled to researchers' capabilities and this may not translate to large, industrial production. Two areas of note are the ethanol extraction methods and hygienic aging, which may have more complications for larger wineries. The resources needed (electricity, labor, equipment, etc.) for certain methods may not be economically viable to scale up to larger production which may limit the integration of these techniques and may limit the production of low-alcohol wine to smaller facilities.

There is still much to learn about winemaking transitioning from producing traditional wine to low-alcohol wine. Reduction of alcohol is a necessary skill due to climatic changes, with the added benefit of consumer preferences and health initiatives, and the industry would do well to invest in the research needed to create desired flavor profiles and efficiencies in the process.

8. Conclusion

Millenia in the making, traditional wine production is filled with cultural and religious significance, but modern trends have moved toward low-alcohol alternatives. Both government and non-government officials have researched the health detriment of alcohol consumption and favor the implementation of economic disincentives to traditional wine to encourage the public to consume less alcohol. With this greater public image of alcohol overconsumption, consumer preferences have steadily trended toward alternative beverages like light beers and low-alcohol wines. Wines with reduced alcohol content also benefit from showcasing the positive health aspects of wine production. Consumer surveys have shown the sale of low-alcohol wine as a functional food may be well received and compensated by the market, which is encouraging to hear for winemakers.

As climate issues and terroir changes favor greater sugar accumulation, vineyard and winemaking practices should be implemented to counter the increased alcohol concentration. Rootstock selection, vigor manipulation, reducing photosynthesis, and timing harvest all help to decrease the fermentable sugars in the must. These practices can also facilitate the development of other flavor compounds, setting up the must for a better flavor profile after harvest. Winemaking practices of yeast selection, premature fermentation arrest, batch blending, and alcohol removal can be implemented to produce a low-alcohol product with better mouthfeel, body, structure, and an overall richer flavor profile.

While there seems for be a promising market for the product, the production of low-alcohol wine has numerous challenges. Grape growers and winemakers will need to adopt certain methods to best prepare and ferment the must to retain a high-quality wine while reducing the alcohol content. Challenges that must be faced by each team include climate change on terroir for the viticulture side and fermentation kinetics, microbial spoilage, and instability for winemakers. Production on a larger scale must also be deliberately planned and monitored to prevent the spoilage with less alcohol to sanitize. Despite its potential, low-alcohol wine requires a thorough strategy to balance the economic gains while maintaining quality and minimizing product loss. More research is needed in the industry to understand new efficiencies or eliminate certain issues of the production of low-alcohol wine.

Author details

Komal Sekhon and Qun Sun*
California State University, Fresno, Fresno, CA, USA

*Address all correspondence to: qsun@csufresno.edu

References

[1] McGovern PE, Fleming SJ, Katz SH. The Origins and Ancient History of Wine: Food and Nutrition in History and Anthropology. London, England: Routledge; 2003

[2] Barnard H, Dooley AN, Areshian G, Gasparyan B, Faull KF. Chemical evidence for wine production around 4000 BCE in the late chalcolithic near eastern highlands. Journal of Archaeological Science. 2011;**38**(5):977-984

[3] McGovern PE, Mirzoian A, Hall GR. Ancient Egyptian herbal wines. Proceedings of the National Academy of Sciences. 2009;**106**(18):7361-7366

[4] Tattersall I, DeSalle R. A Natural History of Wine. New Haven, CT, USA: Yale University Press; 2015

[5] Harrison C. The History of Toasting [Internet]. Englewood, CO, USA: Toastmasters International; 2023. Available from: https://www.toastmasters.org/magazine/articles/the-history-of-toasting

[6] Finucane FM, Campbell N. Sobering evidence that higher taxes will mitigate alcohol-related cancer harms. Lancet Regional Health – Europe. 2021;**11**:100246

[7] Anderson K. Excise taxes on wines, beers and spirits: An updated international comparison. In: The International Economics of Wine. Adelaide, South Australia, Australia: University of Adelaide; 2014. pp. 461-477

[8] Mueller S, Lockshin L, Louviere JJ. Alcohol in moderation: Market potential for low alcohol wine before and after excise tax increase. In: 6th AWBR International Conference, 9-10 June 2011. Bordeaux Management School – BEM – France; 2011. pp. 9-10

[9] Smith S. Economic issues in alcohol taxation. In: Theory and Practice of Excise Taxation: Smoking, Drinking, Gambling, Polluting, and Driving. New York, NY, USA: Oxford University Press Inc.; 2005. pp. 56-83

[10] Bucher T, Deroover K, Stockley C. Low-alcohol wine: A narrative review on consumer perception and behaviour. Beverages. 2018;**4**(4):82

[11] Saliba AJ, Ovington LA, Moran CC. Consumer demand for low-alcohol wine in an Australian sample. International Journal of Wine Research. 2013;**5**:1-8

[12] Masson J, Aurier P, d'hauteville F. Effects of non-sensory cues on perceived quality: The case of low-alcohol wine. International Journal of Wine Business Research. 2008;**20**(3):215-229

[13] Bucher T, Frey E, Wilczynska M, Deroover K, Dohle S. Consumer perception and behaviour related to low-alcohol wine: Do people overcompensate? Public Health Nutrition. 2020;**23**(11):1939-1947

[14] Rehm J, Shield KD, Weiderpass E. Alcohol consumption. A leading risk factor for cancer. Chemico-Biological Interactions. 2020;**331**:109280

[15] Kilian C, Rovira P, Neufeld M, Ferreira-Borges C, Rumgay H, Soerjomataram I, et al. Modelling the impact of increased alcohol taxation on alcohol-attributable cancers in the WHO European region. Lancet Regional Health – Europe. 2021;**11**:100225

[16] Larsson SC, Burgess S, Mason AM, Michaëlsson K. Alcohol consumption and cardiovascular disease. Circ: Genomic and Precision Medicine. 2020;**13**(3):e002814

[17] Bucher T, Deroover K, Stockley C. Production and marketing of low-alcohol wine. In: Morata A, Loira I, editors. Advances in Grape and Wine Biotechnology [Internet]. Rijeka: IntechOpen; 2019. DOI: 10.5772/intechopen.87025

[18] Barreiro-Hurlé J, Colombo S, Cantos-Villar E. Is there a market for functional wines? Consumer preferences and willingness to pay for resveratrol-enriched red wine. Food Quality and Preference. 2008;**19**(4):360-371

[19] Vasiljevic M, Couturier DL, Frings D, Moss AC, Albery IP, Marteau TM. Impact of lower strength alcohol labeling on consumption: A randomized controlled trial. Health Psychology. 2018;**37**(7):658

[20] Provencher V, Polivy J, Herman CP. Perceived healthiness of food. If it's healthy, you can eat more! Appetite. 2009;**52**(2):340-344

[21] Van Leeuwen C, Seguin G. The concept of terroir in viticulture. Journal of Wine Research. 2006;**17**(1):1-10

[22] Spellman G. Wine, weather and climate. Weather. 1999;**54**(8):230-239

[23] Kliewer W. Effect of high temperatures during the bloom-set period on fruit-set, ovule fertility, and berry growth of several grape cultivars. American Journal of Enology and Viticulture. 1977;**28**(4):215-222

[24] Webb L, Watt A, Hill T, Whiting J, Wigg F, Dunn G, et al. Extreme Heat: Managing Grapevine Response. GWRDC Melbourne, Australia: University of Melbourne; 2009

[25] Jordão AM, Vilela A, Cosme F. From sugar of grape to alcohol of wine: Sensorial impact of alcohol in wine. Beverages. 2015;**1**(4):292-310

[26] De Orduna RM. Climate change associated effects on grape and wine quality and production. Food Research International. 2010;**43**(7):1844-1855

[27] Winkler AJ. General Viticulture. London, England: University of California Press; 1974

[28] Shabam PL. The Limitations of the Winkler Index [Internet]. Wine Business Analytics. 2019. Available from: https://winesvinesanalytics.com/columns_article/208245

[29] Gris EF, Burin VM, Brighenti E, Vieira H, Bordignon Luiz MT. Phenology and ripening of *Vitis vinifera* L. grape varieties in São Joaquim, southern Brazil: A new south American wine growing region. Ciencia e Investigación Agraria. 2010;**37**(2):61-75

[30] McKenry MV, Luvisi D, Anwar SA, Schrader P, Kaku S. Eight-year nematode study from uniformly designed rootstock trials in fifteen table grape vineyards. American Journal of Enology and Viticulture. 2004;**55**(3):218

[31] Novello V, Laura de P. Viticultural strategy to reduce alcohol levels in wine. In: Alcohol Level Reduction in Wine-Oenoviti International Network. France: Institut des Sciences de la Vigne et du Vin Faculté d'œnologie - Université Bordeaux Segalen, Villenave-d'Ornon; 2013. pp. 3-8

[32] De Toda FM, Sancha JC, Balda P. Reducing the sugar and pH of the grape (*Vitis vinifera* L. cvs. 'Grenache' and 'Tempranillo') through a single

shoot trimming. South African Journal of Enology and Viticulture. 2013;**34**(2):246-251

[33] Stoll M, Scheidweiler M, Blank M, Schultz H. Possibilities to reduce the velocity of berry maturation through various leaf area to fruit ratio modifications in *Vitis vinifera* L. Riesling. Progrès Agricole et Viticole. 2010;**127**:68-71

[34] Del Amor FM, Cuadra-Crespo P, Walker DJ, Cámara JM, Madrid R. Effect of foliar application of antitranspirant on photosynthesis and water relations of pepper plants under different levels of CO_2 and water stress. Journal of Plant Physiology. 2010;**167**(15):1232-1238

[35] Bindon K, Holt H, Williamson PO, Varela C, Herderich M, Francis IL. Relationships between harvest time and wine composition in *Vitis vinifera* L. cv. Cabernet Sauvignon 2. Wine sensory properties and consumer preference. Food Chemistry. 2014;**154**:90-101

[36] Suárez-Lepe J, Morata A. New trends in yeast selection for winemaking. Trends in Food Science and Technology. 2012;**23**(1):39-50

[37] Varela C, Kutyna D, Solomon M, Black C, Borneman A, Henschke P, et al. Evaluation of gene modification strategies for the development of low-alcohol-wine yeasts. Applied and Environmental Microbiology. 2012;**78**(17):6068-6077

[38] Huang J, Wang Y, Ren Y, Wang X, Li H, Liu Z, et al. Effect of inoculation method on the quality and nutritional characteristics of low-alcohol kiwi wine. LWT. 2022;**156**:113049

[39] Methner Y, Hutzler M, Zarnkow M, Prowald A, Endres F, Jacob F. Investigation of non-Saccharomyces yeast strains for their suitability for the production of non-alcoholic beers with novel flavor profiles. Journal of the American Society of Brewing Chemists. 2022;**80**(4):341-355

[40] Mellor DD, Hanna-Khalil B, Carson R. A review of the potential health benefits of low alcohol and alcohol-free beer: Effects of ingredients and craft brewing processes on potentially bioactive metabolites. Beverages [Internet]. 2020;**6**(2):25. Available from: https://www.mdpi.com/2306-5710/6/2/25

[41] Butzke CE. Survey of yeast assimilable nitrogen status in musts from California, Oregon, and Washington. American Journal of Enology and Viticulture. 1998;**49**(2):220-224

[42] Healthy Wine Fermentations – Understanding YAN and the Nutrient Needs of Your Juice [Internet] 2023. Available from: https://blog.hannainst.com/wine_fermentation_yan_yeast_juice

[43] Erten H, Campbell I. The production of low-alcohol wines by aerobic yeasts. Journal of the Institute of Brewing. 1953;**59**(3):207-215

[44] Blateyron L, Sablayrolles JM. Stuck and slow fermentations in enology: Statistical study of causes and effectiveness of combined additions of oxygen and diammonium phosphate. Journal of Bioscience and Bioengineering. 2001;**91**(2):184-189

[45] Torija MJ, Rozes N, Poblet M, Guillamón JM, Mas A. Effects of fermentation temperature on the strain population of *Saccharomyces cerevisiae*. International Journal of Food Microbiology. 2003;**80**(1):47-53

[46] Edison T. 4 Ways to Stop Wine Fermentation [Ultimate Guide]

[Internet]. 2020. Available from: https://wineturtle.com/how-to-stop-fermentation-in-winemaking/

[47] Aguera E, Bes M, Roy A, Camarasa C, Sablayrolles JM. Partial removal of ethanol during fermentation to obtain reduced-alcohol wines. American Journal of Enology and Viticulture. 2010;**61**(1):53

[48] Lafon-Lafourcade S, Geneix C, Ribéreau-Gayon P. Inhibition of alcoholic fermentation of grape must by fatty acids produced by yeasts and their elimination by yeast ghosts. Applied and Environmental Microbiology. 1984;**47**(6):1246-1249

[49] Schmidtke LM, Blackman JW, Agboola SO. Production technologies for reduced alcoholic wines. Journal of Food Science. 2012;**77**(1):R25-R41

[50] Dealcoholisation of wines. II.3.5.16. International Code of Oenological Practices. Dijon, France: International Organization of Vine and Wine; 2023

[51] Contreras A, Hidalgo C, Henschke PA, Chambers PJ, Curtin C, Varela C. Evaluation of *non-saccharomyces* yeasts for the reduction of alcohol content in wine. Applied and Environmental Microbiology. 2014;**80**(5):1670-1678

[52] Gobbi M, De Vero L, Solieri L, Comitini F, Oro L, Giudici P, et al. Fermentative aptitude of *non-Saccharomyces* wine yeast for reduction in the ethanol content in wine. European Food Research & Technology. 2014;**239**(1):41-48

[53] Varela C, Barker A, Tran T, Borneman A, Curtin C. Sensory profile and volatile aroma composition of reduced alcohol Merlot wines fermented with *Metschnikowia pulcherrima* and *Saccharomyces uvarum*. International Journal of Food Microbiology. 2017;**252**:1-9

[54] Loira I, Morata A, González C, Suárez-Lepe JA. Selection of glycolytically inefficient yeasts for reducing the alcohol content of wines from hot regions. Food and Bioprocess Technology. 2012;**5**(7):2787-2796

[55] Salgado CM, Fernández-Fernández E, Palacio L, Hernández A, Prádanos P. Alcohol reduction in red and white wines by nanofiltration of musts before fermentation. Food and Bioproducts Processing. 2015;**96**:285-295

[56] Salgado CM, Fernández-Fernández E, Palacio L, Carmona FJ, Hernández A, Prádanos P. Application of pervaporation and nanofiltration membrane processes for the elaboration of full flavored low alcohol white wines. Food and Bioproducts Processing. 2017;**101**:11-21

[57] Belisario-Sánchez YY, Taboada-Rodríguez A, Marín-Iniesta F, López-Gómez A. Dealcoholized wines by spinning cone column distillation: Phenolic compounds and antioxidant activity measured by the 1,1-diphenyl-2-picrylhydrazyl method. Journal of Agricultural and Food Chemistry. 2009;**57**(15):6770-6778

[58] Liguori L, Albanese D, Crescitelli A, Di Matteo M, Russo P. Impact of dealcoholization on quality properties in white wine at various alcohol content levels. Journal of Food Science and Technology. 2019;**56**(8):3707-3720

[59] Russo P, Liguori L, Corona O, Albanese D, Di Matteo M, Cinquanta L. Combined membrane process for dealcoholization of wines: Osmotic distillation and reverse osmosis. Chemical Engineering Transactions. 2019;**75**:7-12

[60] Motta S, Guaita M, Petrozziello M, Ciambotti A, Panero L, Solomita M, et al. Comparison of the physicochemical and volatile composition of wine fractions obtained by two different dealcoholization techniques. Food Chemistry. 2017;**221**:1-10

[61] Pham DT, Stockdale VJ, Jeffery DW, Wollan D, Wilkinson KL. Compositional consequences of partial dealcoholization of red wine by reverse osmosis-evaporative perstraction. Molecules. 2019;**24**(7):1404

[62] Pham DT, Ristic R, Stockdale VJ, Jeffery DW, Tuke J, Wilkinson K. Influence of partial dealcoholization on the composition and sensory properties of cabernet sauvignon wines. Food Chemistry. 2020;**325**:126869

[63] Lassen AD, Lehmann C, Andersen EW, Werther MN, Thorsen AV, Trolle E, et al. Gender differences in purchase intentions and reasons for meal selection among fast food customers – Opportunities for healthier and more sustainable fast food. Food Quality and Preference. 2016;**47**:123-129

[64] Bus A, Worsley A. Consumers' sensory and nutritional perceptions of three types of milk. Public Health Nutrition. 2003;**6**(2):201-208

[65] Samoggia A, Riedel B. Consumers' Perceptions of Coffee Health Benefits and Motives for Coffee Consumption and Purchasing - PMC [Internet]. 2019. Available from: https://www.ncbi.nlm.nih.gov/pmc/articles/PMC6471209/

[66] Varela C, Dry PR, Kutyna DR, Francis IL, Henschke PA, Curtin CD, et al. Strategies for reducing alcohol concentration in wine. Australian Journal of Grape and Wine Research. 2015;**21**(S1):670-679

[67] Jolly NP, Varela C, Pretorius IS. Not your ordinary yeast: Non-saccharomyces yeasts in wine production uncovered. FEMS Yeast Research. 2014;**14**(2):215-237

[68] Puligundla P, Pyun YR, Mok C. Pulsed electric field (PEF) technology for microbial inactivation in low-alcohol red wine. Food Science and Biotechnology. 2018;**27**:1691-1696

[69] Bartowsky EJ, Henschke PA. Acetic acid bacteria spoilage of bottled red wine—A review. International Journal of Food Microbiology. 2008;**125**(1):60-70

[70] Hranilovic A, Gambetta JM, Jeffery DW, Grbin PR, Jiranek V. Lower-alcohol wines produced by *Metschnikowia pulcherrima and Saccharomyces cerevisiae* co-fermentations: The effect of sequential inoculation timing. International Journal of Food Microbiology. 2020;**329**:108651

Chapter 5

Trends in Reducing the Effects of Global Warming: Applications of Reverse Osmosis to Obtain Sparkling Wines with Moderate Alcohol Concentrations

Camelia Elena Luchian, Elena Cristina Scutarașu, Lucia Cintia Colibaba, Mihai Cristian Focea and Valeriu Cotea

Abstract

Sparkling wine can undoubtedly conquer the most demanding tastes due to its qualities, uniquely boosted by carbon dioxide. The quality and characteristics of sparkling wines, their stability, and sensory properties largely depend on the physical–chemical composition of the grapes and the base wine, the production technology applied, the environmental conditions, etc. Several techniques can be implemented to produce low-alcohol wines, and reverse osmosis is a procedure that has been successfully used in recent times to reduce the alcohol concentration while having a low negative impact on the composition of the wine under certain conditions. In the current circumstances of changing climatic conditions in areas with a tradition of producing sparkling wines, it is necessary to identify methods of keeping the alcoholic concentration of the base wine at a moderate level to not change the composition of sparkling wines in relation to consumer preferences. This chapter looks at the effects of reverse osmosis and the implications of inoculated yeasts on sparkling wine quality.

Keywords: osmosis, organic acids, volatile compounds, sensory characteristics, low-alcohol wines

1. Introduction

Sparkling wines are considered special due to their high carbon dioxide concentrations, which explains why they are consumed occasionally and for special events. Such wines were obtained in the Champagne region around Reims, an area generally exploited for pastures and cereal cultivation [1]. The Montagne de Reims wine-growing area was favorable to the cultivation of red grapes, while the region of Aÿ-Champagne for white grapes. Dom Pérignon (1638–1715), a Benedictine monk from the Monastery of Hautvillers in Épernay, was the first who obtained, around

1670, an effervescent drink called Champagne, after the name of the wine region of origin. This beverage, initially accidentally produced through the re-fermentation of wine, had a moderate alcohol concentration (about 10% vol. alc.) and a pressure in a bottle of 1.5–2 atmospheres. The sparkling wine was originally produced from red grapes and had a rosé color. Dom Pérignon laid the foundations for what became known as gentle pressing, which allowed minimal contact of the juice with the solids. The varieties cultivated in the Champagne region in those days are not known in detail, but of the 80 varieties, Pinot Noir and Pinot Gris were predominant. With the invasion of phylloxera, most areas cultivated with wines were destroyed, and the existing varieties were subsequently replanted. Although the beverage was originally obtained from seven *Vitis vinifera* varieties, nowadays, it can only be produced from Pinot Noir, Pinot Meunier, and Chardonnay [1]. Also, the name *Champagne* can only be used in the eponymous region. While most sparkling wines are made from a blend of grape varieties, a 'Blanc de Blancs' comes exclusively from Chardonnay, while a 'Blanc de Noirs' comes from black varieties exclusively (usually only Pinot Noir).

Winemakers are seeking ways to lower this crucial parameter for the consumer due to the rise in the alcoholic concentration of wines made from raw materials that grow under conditions of rising annual temperatures and climate change. The techniques employed in this context, including reverse osmosis, usually use membranes, which filter a mixture of compounds with the smallest molecule and then reassemble the obtained solutions to obtain a product with predetermined characteristics. A base wine that has a lower alcoholic concentration is more appropriate for use as, during the secondary fermentation, this characteristic will likely increase. Therefore, the consumer can continue to appreciate an equilibrated sparkling wine [1].

The production of sparkling wines in Romania started with Ion Ionescu de la Brad, a renowned nineteenth-century agronomist who created this beverage for Mihail Sturdza, a prominent Romanian personality of the time [1]. The production and consumption of sparkling wine have seen an upward trend in the last decade, with consumption shifting from mainly special occasions to less traditional contexts [2].

Among the most famous sparkling wines are Cava (produced in Spain), Espumante (Portugal), Sekt (Austria and Germany), Pezsgo (Hungary), and Shampanskoye (Russia). The grape varieties most frequently used to produce sparkling wines globally are Macabeo, Parellada and Xarello (Cava), Pinot Noir, Chardonnay, Pinot Gris, Aligoté, Pinot Blanc, Grolleau, Cabernet Sauvignon, and Chenin Blanc [3].

2. Production of sparkling wines

The fact that sparkling wines are characterized by high concentrations of carbon dioxide earns them the name of special wines. The classification of these beverages is done according to the type and nature of carbon dioxide and the level of pressure in the bottle. Therefore, according to the Organization of Wine and Vine (OIV) regulations, more categories can be found as sparkling, pétillant wines, and pearl (carbonated and lightly carbonated) wines. According to the pressure inside the bottle, there are wines with relatively high pressure (minimum 3 bars) – sparkling and pearl wines; wines with relatively low pressure (between 1 and 1.5 bars): pétillant and lightly carbonated wines. Of these, sparkling wine is obtained through the secondary fermentation of base wine, and carbon dioxide is of endogenous origin only. It is generally sold with an overpressure of 3 bar minimum at 20°C. Compared to sparkling wines, péttilant wines usually have an overpressure of 1–2.5 bar at 20°C. Pearl wines have an overpressure of over 3 bar at 20°C, while the carbon dioxide is of exogenous or partially exogenous origin [1].

2.1 Grape varieties used to obtain sparkling wines

To produce sparkling wines, either white or red base wines can be used. When red grape varieties are used, they are vinified as white wines to obtain the *blanc de noirs*. Necessarily, the selection of the grape variety takes into consideration the pedo-climatic conditions of the region, the productivity of the grape variety, and the desired distinctive character of the wine. The existing sparkling wines tend to use the following varieties: Champagne – Chardonnay, Pinot Noir and Pinot Meunier; Cava – Macabeo, Xarel lo, Parellada, and Chardonnay; Talento – Chardonnay, Pinot Nero, and Pinot Bianco; Asti – Muscato Bianco; Lambrusco – Lambrusco Bianco and Lambrusco Nero; Pinotage – a cross between Pinot Noir and Cinsault; Sekt-Riesling, Silvaner, Pinot Blanc, Pinot Noir, and Pinot Gris. Various clones can also influence the quality of sparkling wines. For example, the Chardonnay clone, VCR10, is recommended in the production of base wine for Cava because of its high acidity, which is particularly suitable for producing sparkling wines. Since the phylloxera crisis (late nineteenth and early twentieth centuries), new rootstocks have emerged from a cross between French and American strains. They were selected according to their degree of adaptation to pedological conditions and European varieties. For instance, 41B is adapted to cretaceous soils and continues to be the most used in Champagne. SO_4 rootstock is adapted to moderately calcareous soils, and 3309C is the strain of choice for slightly calcareous soils [4–6].

2.2 Grape processing

In order to obtain high-quality musts, appropriate conditions during the harvesting and processing of the grapes have to be in place. The harvesting process is extremely important when it comes to quality. It is therefore necessary to harvest the grapes manually and collect them in small buckets or crates to avoid crushing. The fruit loads are shipped immediately for fast processing to avoid the onset of fermentation processes. Sorting the bunches affected by gray rot (*Botrytis cinerea* spp.), which can affect the foaming capacity of the sparkling wine, is also to be considered. The grapes are then promptly but gently pressed, without being previously crushed, at pressures between 1.5 and 2 bar to avoid oxidation. Next, maceration takes place to extract flavor and color compounds, and the process continues with sulfitation of the resulting must (to avoid spontaneous fermentations) in doses ranging from 3 g/hL to 8 g/hL, depending on pressing fraction and crop health. Treatment with pectolytic enzymes and inoculation of *Saccharomyces cerevisiae* starter cultures follows. Musts with turbidity values of 200 NTU to 400 NTU are clarified by centrifugation or static decantation (18–24 hours at 6–15°C), and finishing agents can be added (bentonite or casein). During static settling, some must proteins, pectin as well as mineral cations and phenolic substances are removed [4–6]. The must is further subjected to the fermentation process in order to obtain a base wine with low-alcohol content [7–10].

2.3 Primary fermentation: Obtaining the base wine

The technology for sparkling wines initially requires making a base wine, to be followed by initiation of the second fermentation in the bottle (*Prise de mousse*). After clarification of the must, alcoholic fermentation occurs as a result of inoculation

(10–15 g/hL) with selected strains of *Saccharomyces cerevisiae cv bayanus*. From a biochemical point of view, alcoholic fermentation consists of the anaerobic transformation of sugars (mainly glucose and fructose) under the action of yeasts into ethyl alcohol, carbon dioxide, and other secondary products. However, grape must be a nonsterile substrate that contains several types of yeasts and bacteria that can grow and consequently affect the composition and final quality of the wine. The presence of different yeasts in the must depends on several factors, such as the grape variety, the degree of fruit ripening, the application of pesticide treatments, the degree of fungi development, the climatic conditions, as well as the vine culture technology [11].

Alcoholic fermentation usually takes place in stainless steel tanks at temperatures ranging between 15°C and 20°C, with the fermentation process sometimes occurring at temperatures lower than 13°C. Charcoal can be used to even out and lighten the color of wines, though it does affect the foaming capacity and sensory properties of the final product [12, 13]. As the primary alcoholic fermentation progresses, high amounts of alcohol are formed. The influence of ethanol on yeast has been studied extensively [14, 15]. In general, ethanol tolerance is associated with a higher degree of fatty acids unsaturation from cell membranes [16–18]. Further chemical–physical treatments for clarifying, decanting, and filtering the base wine are followed by bottling. Sparkling base wines will generally have an alcoholic strength of 10–11% vol. alc., low residual sugars, and low volatile acidity. On the other hand, the level of organic acids will be high, which results in increased total acidity (between 12 g/L and 18 g/L tartaric acid) [19].

While the acidity level of the base wine is a key factor for the quality of sparkling wine, its values decrease during the production process due to degradation by yeasts and lactic acid bacteria and precipitation of potassium bitartrate [20].

2.4 Malolactic fermentation

Malolactic fermentation of base wines is widely used in the production of sparkling wines to activate an increase in volatile acidity. The conversion of malic acid-dicarboxylic acid into lactic acid-monocarboxylic acid and carbon dioxide, carried out by lactic acid bacteria, lowers the acidity of the wine and increases its pH. This type of fermentation is important for wines produced from grapes grown in cool climates, which tend to have a high content of organic acids (tartaric and malic) and low pH. In addition, malolactic fermentation ensures microbiological stability. Diacetyl and ethyl lactate are formed during the fermentative process, which gives lactic notes in wines. Malolactic fermentation ought to take place before bottling to prevent the subsequent appearance of bacteria that cause unwanted deposits. Malolactic fermentation is carried out by inoculating lactic bacteria (*Oenococcus oeni*). Some producers do not use this type of fermentation, though, and prefer to preserve the freshness and fruity notes of the wine by applying higher amounts of SO_2 (8 g/hL to 10 g/hL) [21, 22]. Difficulties in inducing malolactic fermentation are usually attributed to the cumulative inhibitory effects of low pH, high alcohol, and SO_2 content of the wines. Nevertheless, such problems could also arise from inadequacy or imbalance of some nutrients (free amino acids) found in sparkling wines [23, 24].

2.5 Conditioning and fining of base wine

Upon completion of malolactic fermentation, wines are subjected to conditioning treatments (gravitational settling and sedimentation or centrifugation, fining with bentonite, gelatin, tannin or silica gel, charcoal, etc.) [24, 25]. Spontaneous/

DOI: http://dx.doi.org/10.5772/intechopen.1003034

gravitational settling is done by keeping the wine in the fermentation vessels for a longer time, decanting and sedimenting the yeast deposit, and separating the wine from it. This can be aided by centrifugation insofar as it replaces the filtration operation with cellulose plates or diatomite. Conditioning the base wine by applying fining treatments involves the use of substances such as protein products (gelatin, casein) and tannin. Moreover, while adding bentonite does help to prevent protein and copper casse, it can also negatively impact the foam-forming capacity. In addition, fining can be accelerated by applying alginates, potassium ferrocyanide (for wines with excess iron and copper), synthetic polyamides (polyvinylpyrrolidone, polyvinylpolypyrrolidone – to eliminate compounds susceptible to oxidation), coal (for color improvement), silicic acid (tannin substitute), etc. Fining of wine by filtration is often used as a complementary process to conditioning as well as for final wines. The application of sulfur dioxide provides protection against oxidation. For sparkling wines, low sulfitation is seen as inhibiting the onset of fermentation. Stabilization against tartaric precipitation is frequently done by refrigeration (keeping the base wine at a temperature of −4°C) or by applying agents that inhibit the crystallization of potassium tartrate (metatartaric acid or carboxymethylcellulose). Ion exchange treatment and electrodialysis can also be used. These treatments are to be carried out only after the final base wine is obtained [1].

2.6 Base wine blend

The practice of blending wines from different grape varieties, different origins, and different years (reserve wines) is important for preserving the quality of sparkling wines. Reserve wines are kept for 2–3 years (sometimes on yeast) in tanks, at 12–13°C, and protected from oxygen; reserve wines can also be kept for a longer period (up to ten years) [24]. The aim of the blending stage is to obtain a base wine with well-defined physico-chemical properties and well-balanced sensory characteristics. The quality of the mixture is mainly established through sensory evaluation [1].

2.7 Secondary fermentation and aging on lees

Secondary fermentation in the bottle (*prise de mousse*) can take place by adding the so-called *liqueur de tirage*, a mixture of sucrose (18–25 g/L depending on the desired CO_2 concentration), and yeasts, diammonium phosphates and adjuvants (bentonite and alginate). The yeasts must allow fermentation at lower temperatures in the presence of the alcohol formed during primary fermentation. The base wine for bottling must have specific characteristics such as light color, fruity aroma, low residual sugars, moderate alcohol content (10–11% vol. alc.), low volatile acidity (acetic acid), and total acidity of 12–18 g/L tartaric acid. Once the wine is in, the bottles are closed using a crown cap with a plastic cylinder, which is meant for yeast collection. The bottles are kept in specially equipped rooms for decanting sparkling wine, at low temperatures and dim lighting. After filling the bottles, they are placed horizontally. Secondary fermentation is slow and takes an average of six to eight weeks in constant low-temperature conditions (11–12°C). The time interval between the addition of the *liqueur de tirage* and the *liqueur de dosage* must be at least 15 months. Once the secondary fermentation is complete, the sparkling wine is matured and aged for a while on the yeast deposit accumulated. Depending on the type of sparkling wine and the legislation in the country of origin, aging varies from 9 to 11 months. During the aging on lees period, the sparkling wine acquires specific organoleptic characteristics acquired through the autolytic process of the yeasts and mediated by hydrolytic enzymes; the latter favor the release of polysaccharides, peptides, fatty acids, proteins, and mannoproteins into the sparkling wine. It is a known fact that oxygen may get into the bottle through the cork, and sensory defects may also occur during aging. Exposure of the bottles to light leads to the degradation of methionine and the formation of volatile compounds with sulfur, which in turn brings about unpleasant smells of cauliflower or wet wool [24].

Yeast autolysis is a very slow process. In general, four main phases occur during aging on lees: (I) during secondary fermentation, the level of amino acids and proteins goes down, and peptides are formed; (II) viable and inactivated cells coexist, peptides are degraded, and amino acids and proteins are released; (III) when no viable cells are present, the release of both proteins and peptides predominates; (IV) approximately 9 months after application of the *liqueur de tirage*, a decrease in amino acid concentration occurs. Proteases favor the hydrolysis of lysosomal and cytoplasmic membranes; they increase the porosity of the yeast cell wall and facilitate the release of degraded constituents into the wine. The slow rate of enzyme activity delays the autolytic process. Consequently, sparkling wines are left in contact with the yeast for several months or years to benefit from the positive autolytic effects [26]. Some studies center on accelerating yeast autolysis, while other authors suggest choosing strains with strong autolytic capacity, combining positive and negative strains with killer factor, and administering *liqueur de tirage* made with exhausted yeast [9, 27]. Another option would be the combination of strains with different autolytic capacity [28]. In this respect, exploiting non-*Saccharomyces* yeasts in combination with *Saccharomyces* strains is worth considering [29]. With a view to improving the complexity of the aroma and diversifying the assortment range, interest in non-*Saccharomyces* yeasts for the production of sparkling wine has soared in recent years.

There is very little data regarding the effect of non-*Saccharomyces* on sparkling wine quality [30, 31], and it mostly covers a limited number of non-*Saccharomyces* specie including *Torulaspora delbrueckii*, *Metschnikowia pulcherrima*, *Schizosaccharomyces pombe,* and *Saccharomycodes ludwigii*. The available studies focus on the analysis of amino acids, ammonia, volatile compounds, glycerol, and protein content, all of which have a bearing on the sensory characteristics of sparkling wines. The fermentation process is monitored by analyzing the content of reductive sugars and measuring the pressure by means of an aphrometer [29].

2.8 Yeast selection for the secondary fermentation

The yeasts that perform the secondary fermentation must have a number of additional characteristics compared to those used in the first fermentation of the base wine production. Besides high resistance to ethanol, they must possess a high flocculation capacity, which facilitates removal from the bottle. Flocculation is a distinctive characteristic of yeasts that is frequently found in *Saccharomyces cerevisiae*, unlike *Saccharomyces uvarum*. As a result of the biochemical processes, several compounds are released which significantly influence the characteristics of the wine and its sensory quality. The selection of yeast strains is important for improving the quality of sparkling wines. Given that mannoproteins are among the major compounds released by yeasts during autolysis, the search for strains that can release large amounts of these compounds is of major interest if one aims to improve sparkling wine quality [32].

2.9 Factors that influence yeast autolysis

The biochemical process of autolysis is influenced by pH level, temperature, the presence of ethanol, and also by the nature of the yeast strain. High temperatures, up to 60°C, have been reported to favor autolysis in a wine model system. Molnar et al. [33] reported that the optimum temperature for proteolysis with the *Champenoise* method is between 10 and 12°C. Nunez et al. [34] compared the autolytic capacity of different strains and proposed this as a criterion for yeast selection. The autolytic capacity was evaluated by measuring the amino acids released by the yeast at different temperatures ten days after fermentation. Significant differences were observed in the autolytic capacity of the three strains. It is a fact, therefore, that the yeast strain acts on the amount of nitrogen released into the environment, which could potentially be useful for sparkling wine production [35]. Martinez-Rodriguez et al. [36] suggested that a yeast strain with good autolytic capacity would produce better-quality sparkling wine than yeast with low autolytic capacity. Also, autolytic capacity together with foam analysis, should be used to select yeasts for sparkling wine production. Nunez et al. [34] recently confirmed that the autolytic capacity of yeast is important for sparkling wine quality.

2.10 Riddling and disgorging

During riddling, the bottles are rotated daily for about 15 days until they are perpendicular to the floor. Thus, the yeasts are directed toward the neck of the bottle. In this way, the negative effects of oxygen and biological degradation are avoided. Riddling is not a homogeneous process, as it depends on the type of yeasts present, the variable surface area, and the flocculation characteristics of the yeasts. The addition of both bentonite and the *liqueur de tirage* aims to normalize sedimentation. The concentrated yeast deposit that reaches the neck of the bottle is then removed through the disgorging process. The process involves placing the bottle in low-temperature brine, freezing the neck of the bottle, and removing the frozen yeast deposit. During this process, part of the liquid is lost, only to be refilled with *liqueur de dosage* consisting of cane or beet sugar (between 6 g/L and 50 g/L sucrose) and antioxidant substances – SO_2, citric or ascorbic acid. During this operation, the pressure in the bottle decreases due to the loss of carbon dioxide. Care must be taken for the dosage not to raise the alcoholic strength of the sparkling wine by more than 0.5% vol. alc. [8, 10].

2.11 Corking

After applying the *liqueur de dosage* and filling the bottles with sparkling wine, they are closed with corks (made of cork, synthetic materials – polyethylene, metal with thread). In the case of cork closures, it is necessary to fix them with wire hoods in order to withstand overpressure [1].

3. Methods of producing sparkling wines

The production method is one of the key factors that determine the style and quality of white and rosé sparkling wines. The sparkling wine market has expanded significantly over time, mostly driven by increased global consumer demand. Due to its high added value, the economic impact of sparkling wine is very important, even if production is lower compared to still wines [7]. Current trends highlight a rising consumer interest in sparkling wines, which in turn increases demand on the global market. The type of inoculated yeasts is a fundamental parameter for optimizing the production technology as well as improving the quality of the end-product. According to the regulations and guidelines of the International Organization of Vine and Wine [37], sparkling wines belong to the category of special wines. Sparkling wines differ mainly in terms of production technology, of which the following stand out: the Champenoise method, Cremant (France and Luxembourg), the transfer method, the ancestral method/*méthode ancestrale* (Limoux, Gaillac), Dioise and the Charmat method (in tanks).

The Champenoise method involves obtaining the base wine by completing the alcoholic fermentation process, followed by a secondary fermentation and aging in bottles (**Figure 1**). This method involves the removal of the sediments formed by the riddling and disgorging processes [38]. It is used in France to obtain Champagne, which uses varieties such as Chardonnay, Pinot Noir and Pinot Meunier, Arbanne, Petit Meslier, and Pinot Blanc [7].

Transfer method isobarometric transfer. After fermentation and maturation in bottles, the base wine is cooled and then transferred to a pressure tank by means of an automatic or semi-automatic installation (**Figure 2**). At this stage, dosing of the *liqueur de dosage* can be done directly in the tank or after filtering the sample. Following aging on lees, the wine is filtered and bottled or simply bottled if filtering was carried out previously. All operations are performed in a carbon dioxide atmosphere. This method ensures a more uniform adding of the *liqueur de dosage*, as well as the likely elimination of riddling and disgorging. However, the transfer method is expensive, energy intensive, and poses an increased risk of wine oxidation [1, 38].

The ancestral method was first devised in the vineyards of Limoux and Gaillac (France), but it is generally difficult to control (**Figure 3**). The base wine is obtained by processing whole grapes (mainly from the Mauzac variety) or applying classic white winemaking technology to an incompletely fermented wine. At various stages of the winemaking process, it is essential to halt the fermentation each time it starts to accelerate. As such, refrigeration (down to 0°C), sulfiting, and depletion of nutrients from the yeast (by decantation, fining, filtration, or centrifugation) are repeated as often as necessary to regulate or stop yeast activity. The wine is then filtered and stored at 0°C for 2–3 months. It is at this stage that the second fermentation takes place (2–3 months) in the bottle (at rigorously controlled temperatures) due to the sugars left over after the first fermentation. Then, there is riddling and disgorging without adding the *liqueur de dosage*. This method may also skip the disgorging

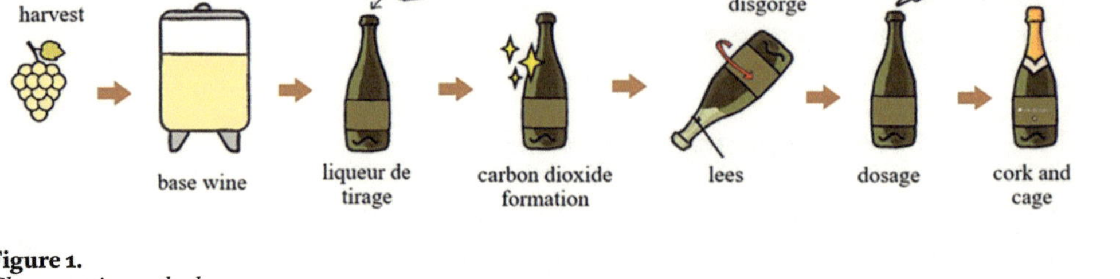

Figure 1.
Champenoise method.

Figure 2.
Transfer method.

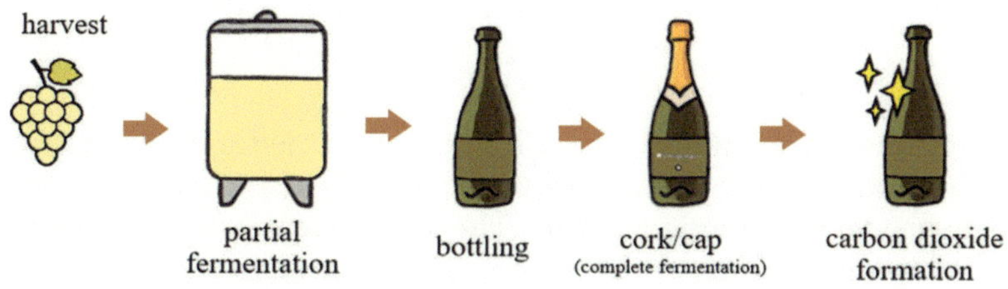

Figure 3.
Ancestral method.

process and keep a low proportion of sediment at the bottom of the bottle. The method is particularly used in France to obtain Cremant, which incorporates such varieties as Chenin Blanc, Chardonnay, Cabernet franc, Pineau D'Aunis, Grolleau, Gamay, Aligoté, Melon, and Sacy [7].

The Dioise method implements the same winemaking principles as *ancestral* to obtain a semi-fermented wine from the Muscat à petits grains variety, filtering and refrigeration at 0°C, and the second fermentation in the bottle. To improve the extraction of flavor compounds, pectinolytic enzymes are administered during grape processing. In this case, the must is characterized by a high turbidity value (1000 to 1500 NTU) that needs to be settled by means of the flotation technique. The second fermentation is halted by applying low temperatures and then transferring the content of the bottles to a stainless steel tank to be kept at an isobarometric pressure of CO_2 to avoid loss of carbon dioxide (similar to the transfer method). After filtration, the wine is bottled by applying the isobarometric principle. The final alcoholic concentration of these beverages is generally around 7.5% vol. alc., with 40–50 g/L of residual sugars [21].

The Charmat method (fermentation in tanks). Unlike the *Champenoise* method, with the Charmat method, the second fermentation occurs in stainless steel tanks. This is a simpler technique that is low-cost compared to the others. Yeasts and sugars are added, and the wine is kept at a temperature of 20–25°C. The second fermentation usually lasts 10 days and is stopped by the application of sulfur dioxide and by refrigerating the wine to −2°C. Having been cold-stabilized at −5°C for several days, the wine is filtered at a low temperature and then bottled using the isobarometric principle. A disadvantage of this method is that it does not mature the wine on the yeast deposit (**Figure 4**) [24].

The method is frequently used to obtain sparkling wines in France (from Cabernet Sauvignon, Chenin Blanc, Cabernet Franc, Sauvignon Blanc, Gamay, Chardonnay, Grollo, Pinot D'Aunis, Pinot Noir, Malbec) and Germany (from Pinot Blanc, Silvaner, Riesling, Pinot Gris and Pinot Noir) [7].

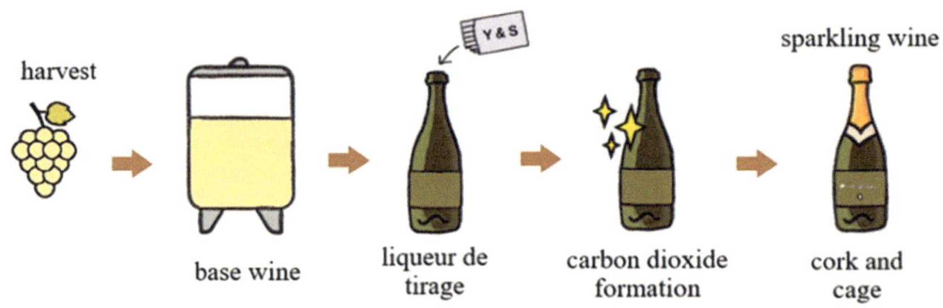

Figure 4.
Charmat method.

Continuous flow fermentation. The continuous flow method, the so-called Russian method, was created in response to the need to manufacture sparkling wines quickly, cheaply, and with minimal effort. The technological steps are the same as for the tank-based fermentation method, except that maturation takes place in a unique bioreactor-style tank, with specific processes ranging from 3 to 5 years. As a result, this method's application necessitates the existence of suitable facilities outfitted with high-capacity pressure tanks, which is a disadvantage to its application [1].

The production of red or rosé sparkling wines is relatively low due to the choice of the moment for harvesting the grapes. In order to be suitable for base wine, red grapes often do not reach the appropriate phenolic maturity that gives the wine organoleptic complexity [28].

4. Yeast selection in the technology of sparkling wine production

The selection of starter cultures is generally based on their oenological properties, which reflect the degree of adaptation to the environmental characteristics specific to must and sparkling wine. The properties of yeasts for the first and second fermentation are different. Nevertheless, it is essential to identify the starter cultures that tolerate the specific characteristics of the must and the base wine. The *Saccharomyces* genus belongs to the *Ascomycota* division and includes seven species, among which *Saccharomyces cerevisiae*. It adapts well to conditions of low pH, high concentration of sugars (osmotic stress), progressive increase in ethanol, and depletion of the corresponding nutrients. *Saccharomyces cerevisiae* is considered the most important yeast species in the winemaking process and is used industrially to carry out high-quality fermentation processes. *Saccharomyces cerevisiae* strains for the secondary fermentation of the base wine usually show additional physiological and technological characteristics when compared to those proposed for the 'primary' starter cultures. The objective is to fully convert the sugars added to the base wine, via the *liqueur de dosage*, into ethanol and carbon dioxide. Thus, it is necessary that the strains can activate in an environment of at least 10–12% vol. alc., low pH (2.9–3.2), tolerate low temperatures (10–15°C), total SO_2 concentrations of 50–80 mg/L, high total acidity (12–18 g/L tartaric acid), low volatile acidity (0.2–0.4 g/L), high pressure (5–6 bar) and glycerol content of 5–20 g/L. On top of that, some authors highlight the importance of the flocculation and autolytic capacities of yeasts during secondary fermentation. Flocculation is a special characteristic of yeasts that allows for the settling of musts during fermentation. This physiological feature can be described as a natural ability of yeast strains to form compact microbial biomass, which settles at the bottom of the fermentation vessel and, following foam development, facilitates the disgorging of the deposit that accumulates at the neck of the bottle. Not least important, yeast flocculation appears to be associated with improved ester production.

Interspecific hybrid strains are also suitable for the production of sparkling wine re-fermented in the bottle; they are obtained through the appropriate selection of flocculant strains of *Saccharomyces cerevisiae* and nonflocculant strains of *Saccharomyces bayanus.* Interspecific hybrids are able to ferment at a temperature range between 6°C and 36°C. The autolytic capacity of yeast is inherent to another selective character of yeast: the killer phenotype, which releases toxic proteins. Under oenological conditions, acceleration of the autolytic process is pursued through the appropriate mix of killer and killer-sensitive yeast strains and the use of mutant yeast strains with autolytic characteristics. Mutagenesis of *Saccharomyces cerevisiae* strains induces accelerated release of proteins, amino acids, and polysaccharides. Sparkling wines obtained from selected mutant strains show better foaming properties than those inoculated with nonmutant strains. The autolytic capacity of yeasts is closely related to the release of volatile compounds in sparkling wine and their impact on the sensory profile. Research analyzes a specific mannoprotein (PAU5) synthesized by certain *Saccharomyces cerevisiae* strains. This mannoprotein, encoded by the seripauperin gene PAU5, directly diminishes the undesirable phenomenon caused by the agitation of sparkling wines and, as such, has foam-stabilizing properties. Excessive spontaneous foaming triggered by the release of pressure upon opening the bottle causes important economic losses. This can be caused either by contaminated raw material (for example, the presence of the bacteria *Botrytis cinerea* spp.) or by noncompliance with the production process. Following the secondary fermentation of sparkling wine, *Saccharomyces cerevisiae* yeasts are able to produce indole even when the viability of the culture is very low. The proportion of CO_2 released by yeasts and dissolved in the liquid phase during the fermentation process is an additional parameter for the selection of yeast cultures, one of fundamental importance for the organoleptic perception of the end-product [10, 28]. The sensory properties of sparkling wine are influenced by various factors, such as the processing technology, the grape varieties, the chemical composition and structure of the base wine, the selected yeast strains, and the aging time spent on lees. However, the sensory profile of sparkling wines depends more on the autolytic release of volatile compounds during the above-mentioned phase since aroma is one of the most relevant indicators of sparkling wine quality. Primary (prefermentative) aromas are characteristic of the grape variety used, secondary (fermentative) aromas are released following the metabolism of yeasts during fermentation, and tertiary (postfermentative) aromas are formed during aging. Cotea et al. [39] studied the influence of four commercial yeasts on the volatile profile of sparkling wines and found that the enrichment of aromatic compounds is a specific characteristic of inoculated yeast strains. Sensory evaluation and chemical analysis of volatile compounds by qualified experts are among the most important techniques for assessing the aroma profile of sparkling wines. At the end of the second fermentation, the long phase of aging on lees begins and with it, the release of intracellular compounds that causes an increase in free amino acids, often deemed as precursors of aromatic compounds in sparkling wine. All along the aging of sparkling wine, the autolysis of yeasts releases a considerable number of volatile molecules (esters, higher alcohols, aldehydes, sulfur compounds, organic acids, etc.), a process closely correlated to the yeast strains that are present. Being an enzymatic hydrolysis of biopolymers, autolysis is a very slow process associated with the death of yeast cells and the release of volatile constituents.

There are studies that focus on analyzing the different compounds released during the autolytic process and exploring additional methods of inducing autolysis in order to influence the aging character in sparkling wine. There are authors who investigate the different compositional changes that occur during the production of sparkling wines and the factors that most influence a range of sensations at the olfactory level. Polysaccharides can influence the viscosity, foaming capacity, and sensory perception of sparkling wines. Mannoproteins are released in sparkling wine during yeast autolysis to prevent protein disorder or crystallization of potassium bitartrate. Alcohols, carbonyl compounds and organic acids are mainly responsible for the burned, plastic, or rancid note. Aldehydes, such as octanal, nonanal, decanal, and some terpenes, enrich sparkling wines with citrus-like aromas. Intense chemical, roasty aromas, floral, vegetable, and fatty notes are discerned in sparkling wines produced with strains of *Saccharomyces cerevisiae* bio-immobilized with filamentous fungi – *Penicillium chrysogenum*. It is therefore clear that not only the strain but also the method of immobilization of the yeasts influences the final sensory characteristics of sparkling wines. The presence of indole and other volatile compounds definitely affects the sensory perception of these products because of the unpleasant plastic flavors. Both *Saccharomyces* and non-*Saccharomyces* yeasts can yield high concentrations of indole during primary fermentation. Extensive research on the selection of indigenous yeasts for use in grape must fermentation has been carried out, but few authors have explored the study of indigenous yeasts in sparkling wines. In the latter case, the selection of indigenous yeasts is much more complex because the base wine, into which the specific strain of *Saccharomyces cerevisiae* is inoculated, is considered an unfavorable environment for the growth of microorganisms due to its high alcohol content and low pH. Indigenous strains of *Saccharomyces cerevisiae* are selected based on several technological criteria (fermentative strength and vigor, SO_2 and ethanol tolerance, and flocculation capacity) and qualitative characteristics (acetic acid content, glycerol, and hydrogen sulfide production) [8, 40]. An interesting property of indigenous strains of *Saccharomyces cerevisiae* used in sparkling wine production is their ability to modulate the phenolic components of the end-product. The appropriate selection of the *Saccharomyces cerevisiae* strain and its proven ability to enhance the varietal properties of grapes are important in the production of a quality sparkling wine with a sensory profile that reflects the typicality of the grape variety [41]. Tufariello et al. [42] studied comparatively the sensory profile of sparkling wines obtained with indigenous strains of *Saccharomyces cerevisiae* (isolated in Salento, Apulia, Italy) and commercial strains DV10® (Lallemand Oenology, Grenaa, Denmark), respectively. For samples obtained with DV10® yeasts, high concentrations of gluconic acid were recorded, which negatively influenced the foaming properties of the end-product; however, this was not the case with the sparkling wines obtained from autochthonous strains of *Saccharomyces cerevisiae.* In addition, the use of selected indigenous strains also led to high proportions of volatile compounds (roses and fruity aromas), low values of volatile acidity, and high glycerol content [43]. Other studies showed a reduced flocculation capacity and low volatile acidity of indigenous strains isolated both in Apulia and in the Lombardy area (Italy), compared to commercial ones. In particular, indigenous strains of *Saccharomyces cerevisiae* isolated from grape berries in Apulia (Italy) were tested for tolerance to different stress factors such as pH, ethanol content, and total SO_2 level. By monitoring fermentation at 6°C and 12°C respectively, most native strains produced very low CO_2 values and exhibited low autolytic and flocculant properties, killer activity, resistance to pH 3.5 and tolerance to ethanol concentrations between 6 and 12% vol. alc., as well as at various concentrations of total SO_2 (100, 150 and 200 mg/L). Also, the volatile acidity of these samples was low, while the pressure in the bottle reached about 5 bar. In the study by Vigentini et al. [40], native strains isolated from the Oltrepò Pavese region (Italy) released low concentrations of glycerol and high amounts of hydrogen sulfide. They were characterized by increased resistance to a concentration of 12% vol. alc., reduced ability to tolerate high concentrations (300 mg/L) of SO_2, and low volatile acidity. Alfonzo et al. [3] selected indigenous strains of *Saccharomyces cerevisiae* for the secondary fermentation of the base wine obtained from the Grillo variety. The four strains exhibited solid resistance to fermentation and sulfur dioxide, good wine re-fermentation capacity at high total acidity and very low pH, with no off-flavors. Studies have highlighted a high level of genomic diversity within the *Saccharomyces cerevisiae* species.

Saccharomyces yeast strains (*Saccharomyces bayanus, Saccharomyces oviformis Osterwalder*) were used by Bozdogan et al. [44] in both free and immobilized form, the latter after immobilization in alginate beads to enhance the secondary fermentation of the base wines obtained from the Emir and Drimit varieties. Significant differences in amino acid concentrations were detected as a factor of the maturation period and the yeast inoculated. The beneficial effects of non-*Saccharomyces* yeasts (e.g., *Torulaspora delbrueckii*, *Pichia kluyveri*, *Lachancea thermotolerans*, and *Metschnikowia pulcherrima*) on still wines have been showcased extensively in the literature, but few studies research their impact on the quality of sparkling wines. González-Royo et al. [11] tested two strains of *Torulospora delbrueckii* and *Metschnikowia pulcherrima* to produce base wine from the Macabeo variety through successive inoculations with *Saccharomyces cerevisiae*. They found that the sparkling wine obtained using *Metschnikowia pulcherrima* had a longer persistence of the foam and an interesting aromatic profile, with notes of smoke and flowers. Reports show that *Torulaspora delbrueckii* decreases volatile acidity but increases the proportion of glycerol, thus improving the foaming properties of wine due to the autolysis of yeasts [45]. Sparkling wines produced with *Saccharomycodes ludwigii* and *Schizosaccharomyces pombe* display important changes in color characteristics, acidity, volatile profile, and biogenic amines [31]. The sequential inoculation of *Torulaspora delbrueckii* and *Saccharomyces cerevisiae* helps obtain sparkling wines with high protein content and improved foaming properties [46], but also large amounts of ethyl propanoate, isobutyric and butanoic acid, alcohols, and phenols. However, Velázquez et al. [46] propose that only *Torulaspora delbrueckii* yeasts be inoculated under strict conditions (high pressure and alcohol content). These yeasts do not complete the secondary fermentation of the sparkling wine and leave high amounts of reductive sugars, plus low CO_2 production and, consequently, low pressure. It is important to point out that some non-*Saccharomyces* yeasts can have a negative effect on sparkling wines. For example, *Zygosaccharomyces* species overproduce acetic acid.

Following the tendency of consumers to turn to drinks with a lower alcohol concentration, correlated with the preference of the new generation for different organoleptic sensations, Focea [47] evaluated the influence of yeasts (*Saccharomyces* spp.) used in the second fermentation on the quality of white sparkling wines obtained from the Muscat Ottonel variety. The results show that yeasts can influence the final quality of sparkling wines in various ways. Considering the physico-chemical characteristics, the type of inoculated yeasts was found to have a minor, though important, impact on the physico-chemical parameters. The metal composition of wine during fermentation, maturation, and storage is not stable. The analysis of variance reveals a significant influence of the metal content according to the type of inoculated yeasts.

The results identify a higher content in the organic acids analyzed in the case of the base wine compared to the sparkling wines obtained at a later time. No major differences were signaled in the concentration of the main organic acids and pH values. At an early stage, the concentration of organic acids showed minor variations in relation to the type of yeasts inoculated for the second fermentation ($p > 0.05$). Significant changes were found between the two analysis points (6 and 11 months of storage). The yeasts administered for the second fermentation contributed to a significant increase in the concentration of most volatile compounds ($p < 0.05$).

The quantified volatile compounds belonged to different chemical classes including esters, acids, alcohols, and terpenes. In the first category, the presence of ethyl octanoate, ethyl decanoate, ethyl laureate, isopropyl myristate, ethyl palmitate, and ethyl oleate was noted. Their concentrations varied according to the inoculated yeasts. The data showed a significant contribution of the selected yeasts to the enrichment of the volatile fraction of the wines. Regarding organoleptic characteristics, important differences were obtained depending on the yeast product used. Ethyl octanoate and ethyl decanoate were well represented in all variants, which explains the fruity (especially banana, and apple) and floral (elder flower) notes of the samples. In our team studies, the principal component analysis describes the variations in the composition of volatile compounds of sparkling wines produced under the influence of different strains of commercial yeasts. The first factor of the data variability was closely correlated with most of the volatile compounds identified (ethyl octanoate and decanoate, 2-phenethyl acetate, ethyl laurate, hexanoic, octanoic, decanoic and 9-decenoic acid, alcohol isoamyl, 4-octanol, phenylethyl alcohol, linalool L and α-terpineol). As such, these components showed a high correlation with most of the volatile compounds identified in the samples analyzed. The first principal component that explained most of the total variability in the data was highly correlated with isoamyl acetate, ethyl decanoate, ethyl laurate, isoamyl alcohol, and linalool. For the second main component, diethyl succinate and isopropyl myristate showed high and positive values. A positive correlation of 1-heptanol with butyric acid (r close to +1) and a negative correlation with isopropyl myristate (r close to −1) was evidenced. In addition, linalool and ethyl decanoate are positively correlated, while linalool and butyric acid present a negative correlation. Ethyl octanoate, ethyl decanoate, and isoamyl alcohol, the most prevalent volatile substances in the samples analyzed, were also positively correlated. On the other hand, they are negatively correlated with butyric acid and 1-heptanol (positioned in the opposite direction). The variables related to factor 1 allow the differentiation of the samples according to the volatile fraction. From a sensory point of view, the sparkling wines analyzed were characterized by intensely fruity (bananas and apples) and floral (elder flowers) notes associated with high levels of esters (e.g., ethyl octanoate and ethyl decanoate). Regarding concentration, compounds such as isoamyl acetate, ethyl palmitate, 4-octanol or 1-heptanol did not significantly contribute to the sensory profile identified. Samples were described as balanced on the palate, with increased persistence, high acidity (conveying freshness), and good texture. A significant influence of the yeasts on the final sensory profile was noted, with the control sample showing the lowest scores for most of the descriptors followed. The variant obtained with IOC 18–2007™ boasted high acidity and effervescence, a good texture (highest note), and increased persistence. The sample labeled IOC DIVINE™ was highly rated for its intensely fruity notes (apples, melon), while the LEVULIA CRISTAL™ variant stood out for its vegetal, yeasty, and also apple character.

5. Chemical composition of sparkling wines

There are many factors involved in the chemical composition of sparkling wines, such as grape variety, cultivation technology, quality of base wine, type of inoculated yeasts for secondary fermentation, etc. However, the second fermentation and the maturation stage of the yeast sediment are key factors for the final quality of sparkling wines [48].

Water is found in the highest proportions of all existing compounds in sparkling wines (60–85%), with a direct participation in the occurrence of chemical reactions [49].

Alcohols represent, on average, 15% of the total constituents, the predominant ones being ethyl alcohol, 1-propanol, 2-phenylethanol, 2-methyl-1-propanol, etc. They can serve as solvents for some chemical constituents, participate in the formation of compounds such as esters, acetates, and also play an essential role in defining the sensory profile [50].

Acids are found in smaller proportions compared to the first two components, although they show key roles in defining the quality of sparkling wines and have a major impact on the organoleptic profile. Of these, minerals (sulfuric acid and phosphoric acid) and organic acids (tartaric acid, citric acid, malic acid, etc.) stand out [51].

Phenolic compounds can be identified in amounts up to 500 mg/L in white sparkling wines and are of particular interest due to their antioxidant capacity. This category includes phenolic acids (gallic acid, vanillic acid, gentisic acid, syringic acid, etc.), tannins (catechins, procyanidins, and proanthocyanidins), pigments (anthocyanin and flavones), which all engage in numerous redox reactions [48, 50]. Despite the obvious influence of phenolic compounds on the sensory quality of wines, there are not many studies aimed at tracking the changes in these compounds during the technological process. Jeandet et al. [24] notice an intensification of the brick-red color of sparkling wines after 15 months of maturation-aging. This process may be due to the oxidation of phenolic compounds released during autolysis of the yeast, as well as to the action of cytoplasmic (also released during autolysis) and hydrolytic enzymes. In addition, this phenomenon was reduced at the beginning of the autolysis period (between the 12th and 18th month), possibly due to the release of phenolic compounds adsorbed by the yeast. Also, a reduction in resveratrol content during aging on yeast was attributed to the adsorption capacity of the yeast for phenolic compounds.

Aroma compounds can originate in the raw material (primary aromas) or can either be formed during fermentation processes (secondary) or during the maturation-aging period (tertiary). The concentration of these compounds depends on the quality and composition characteristics of the raw material and also on the technology of obtaining sparkling wines. The odorous compounds typical of still and sparkling wines belong to the class of terpenes, phenols, esters (ethyl acetate, isoamyl acetate, benzyl acetate), thiols (4-mercapto-4-methyl-2-pentanone, 3-mercapto-1-hexanol), isoprenoids (geraniol, α-terpineol, linalool), etc. [52]. Most of the early studies related to changes in volatile compounds during the aging of sparkling wines are contradictory. Some authors find an increase in these compounds during aging, while others find a decrease in the concentration of ethyl esters and acetates. This may be accounted for by differences in the experimental conditions as well as the simultaneous degradation and synthesis of volatile compounds that occur during the aging of wine with yeast; as a result, either of these processes predominates at any given moment [48]. Pozo-Bayón et al. [49] showed a major impact of second fermentation and yeast maturation time on the proportion of volatile compounds in sparkling wines. They reported that the proportion of volatile compounds such as hexyl acetate, isopentyl acetate, ethyl butyrate, ethyl octanoate and diethyl succinate can provide information concerning the age of sparkling wines. It has also been suggested that the increased level of C13 norisoprenoids in sparkling wines aged for longer periods of time (over 21 months) may be due to the action of enzymes released during yeast autolysis on the pigments (carotenoids) present in the wine. Of these, vitispirane is considered a marker of the maturation-aging process in sparkling wines [53].

Nitrogen substances are present in wines either in mineral or organic form. These compounds are found in low concentrations in wine and usually come from the raw material. Nitrogen substances play an important role in the development of fermentation processes and foaming capacity [54]. Curioni et al. [55] reported the important antioxidant, surfactant, and antimicrobial role of these compounds. Of all nitrogen substances, amino acids (glycine, valine, proline, phenylalanine, histidine, arginine, etc.) add up to 30% of the total figure in the case of white sparkling wines. Amino acids are food for yeasts during fermentation. During fermentation, the proportion of peptides first increases and then decreases toward the end of the process. The decrease has been attributed to the consumption of peptides by the yeasts and the presence of active acid proteases in the wine [54].

The amino acid fraction of the base wine is the main source of nitrogen during fermentation. These acids also serve as precursors to aromatic compounds that contribute to the characteristics of sparkling wines. Amino acids in sparkling wines come from various sources, such as the grapes used to obtain the base wine, which are not metabolized by yeasts during growth, while others are released by yeasts at the end of fermentation or during autolysis. This may be either because more peptides than amino acids are released during autolysis, or the released amino acids (glutamic acid, arginine, and alanine) are converted by decarboxylation and deamination, which has resulted the reduction of the amino acid fraction [54].

Proteins are important in improving the quality of the effervescence and increasing the stability of the foam. It has also been reported that polysaccharides and mannoproteins play a positive role in enhancing taste perception and that some proteins and peptides released by yeasts may contribute to the sweet taste of wine. In addition, some amino acids, peptides, and nucleotides can contribute to the umami taste and, as such, have been labeled flavor enhancers. It has been found that amino acids and lipids are aroma precursors, and consequently, their release from yeast cells can also contribute to the aromatic complexity of sparkling wines [55].

Sulfur compounds are found in low proportions in sparkling wines, both in inorganic (hydrogen sulfide) and organic form (sulfones, thioalcohols, thioesters, thioethers, etc.). These compounds usually have an unpleasant odor (hatched egg, rotten cabbage or rubber) [50].

Polysaccharides are macromolecules that originate in the raw material, yeasts, bacterial, and fungal contamination of grapes (*Botrytis cinerea* spp.). Arabinose- and galactose-rich polysaccharides such as type II arabinogalactans and arabinans, type I and type II rhamnogalacturonans and homogalacturonans come from grape seeds, while glucans, mannans, and mannoproteins are released from the yeast either during fermentation or by enzymatic action during aging on yeast, by autolysis. Exogenous polysaccharides such as arabic gum and carboxymethyl cellulose can be found in still and sparkling wines and are considered additives. Their action in wine is dependent on their type but also on the concentration in which they are found. Arabinogalactans greatly influence the filtration process, while mannoproteins are more effective in reducing the cloudy appearance of white wines. Rhamnogalacturonans, mannoproteins, and arabinogalactans influence the aggregation of proanthocyanidins differently, which explains their varying effects on wine characteristics. These compounds could well constitute markers for monitoring the autolysis process [56].

Reductive sugars represent the main classification criterion for sparkling wine, as follows: natural brut (maximum 3 g/L), extra brut (maximum 6 g/L), brut (maximum 12 g/L), extra dry (12.01–17 g/L), dry (17.01–32 g/L), demi-dry (32.01 –50 g/L), and sweet (minimum 50.1 g/L). The reductive sugars are either the result of an incomplete or stopped fermentation or the presence of nonfermentable carbohydrates (arabinose and xylose). Alternatively, they can be added during the technological process (application of sucrose, concentrated must, etc.) [51].

In sparkling wines are usually found large amounts of gaseous compounds such as carbon dioxide, sulfur dioxide, and nitrogen [49].

The predominant vitamins in must and wine include the B complex, C, F, and P. They are growth factors for microorganisms (yeasts and bacteria) during the biochemical processes of fermentation. Vitamin content decreases following bentonite application [56].

The presence of mineral substances in sparkling wines depends on the composition of the raw material, the *terroir*, and the technology. The presence of potassium, iron, sodium, copper, calcium, magnesium, zinc, etc. has been noted [49].

6. Implications of the technology applied for the quality of sparkling wines

Most of the studies on sparkling wines focus on methods for improving their quality. Given that sparkling wine composition is the result of many factors, different approaches to quality management have been considered. A general trend is the diversification of the assortment range, and to this purpose, there are studies that look at the potential of different grape varieties. The varieties specific to each geographical region have been found to lend a varietal imprint that gives uniqueness to the product [57]. Along the same lines, research has targeted indigenous yeasts [58], seen as microorganisms that reflect the biodiversity of a certain area, which further supports the idea that indigenous yeast strains can be associated with the *terroir* [59]. However, sparkling wine fermentation is a difficult challenge for indigenous yeasts. Other authors study the influence of secondary fermentation conditions, the selected method, and aging of yeast on the quality of sparkling wines [32].

Ruiz-Moreno et al. [60] investigated the influence of prefermentative maceration and aging on the aroma profile (with emphasis on ester content) of some sparkling wines obtained from the Pedro Ximenez variety. The prefermentative maceration process kept the skin in contact with the must at 10°C for 6 hours. The Champenoise method was applied to obtain the sparkling wines, and the samples were monitored at 3, 6, and 9 months of maturation on the lees. Sparkling wines obtained with prefermentative maceration showed higher contents of branched-chain ethyl esters and cinnamic acids. On the other hand, samples obtained without maceration showed higher levels of ethyl esters of fatty acids and acetates of higher alcohols. The results underscored the impact of prefermentative maceration and aging on the aroma profile of sparkling wines. Prefermentative maceration had a significant bearing on the aroma profile of sparkling wines. Higher levels of branched-chain ethyl esters, cinnamates, fatty acid methyl esters, odd-carbon chain ethyl esters of fatty acids, and various compounds were found in sparkling wines obtained by prefermentative maceration. Ethyl heptanoate, ethyl phenylacetate, methyl hexanoate, and ethyl propanoate stood out as potential volatile markers of prefermentative maceration, while ethyl isovalerate, ethyl isobutyrate, and ethyl 2-methylbutyrate were identified as markers for maturation-aging on lees.

Tofalo et al. [61] focused on the effects of inoculation with *Saccharomyces cerevisiae* (F6789) and *Torulaspora delbrueckii* (TB1) or *Starmerella bacillaris* (SB48) on the quality characteristics of sparkling wines obtained by the Champenoise method. The autolytic properties and the sensory profile of the sparkling wines obtained were also assessed. Secondary fermentation was completed by all mixed and single starter cultures except those driven by *Starmerella bacillaris*. Sparkling wines produced with a yeast mix of *Saccharomyces cerevisiae* (F6789) and *Starmerella bacillaris* (SB48) presented the highest amounts of glycerol (6.51 g/L) and the best autolytic potential (81.98 mg leucine /L). The lowest value for the autolytic indicator was observed for samples obtained only with *Saccharomyces cerevisiae* – F6789 (53.96 mg leucine/L). The sparkling wines exhibited different aromatic and sensory profiles, showing higher concentrations of esters in the variants produced only with *Saccharomyces cerevisiae* – F6789 (88.09 mg/L) followed by those obtained with a mixture of *Saccharomyces cerevisiae* + *Torulaspora delbrueckii* (87.20 mg/L), and finally, the lowest values were recorded for samples inoculated with a mix of *Saccharomyces cerevisiae* and *Starmerella bacillaris* (81.93 mg/L). The ester content was seen to decrease with time, which may be related to adsorption on yeast and chemical hydrolysis. The highest concentrations of higher alcohols were found in sparkling wines produced with *Saccharomyces cerevisiae* + *Torulaspora delbrueckii* (27.50 mg/L). The sparkling wines obtained with *Saccharomyces cerevisiae* and *Starmerella bacillaris* had distinctive spicy notes, aroma of toasted bread, freshness, and floral smell.

Pérez-Magariño et al. [62] studied the volatile composition of some white and *rosé* base wines made from different native Spanish grape varieties (Verdejo, Viura, Malvasía, Albarín, Godello, Prieto Picudo, and Garnacha), as well as the quality of sparkling wines obtained by the Champenoise method. To that purpose, the number of amino acids and biogenic amines were analyzed, and the results showed that both the base and sparkling wines obtained from the Albarín, Verdejo, Godello, and Prieto Picudo varieties presented the richest volatile profile, boasting high concentrations of ethyl esters and acetates of alcohols which contribute to the definition of the fruity aroma of the wines. During aging of yeast, an increase in branched ethyl esters, ethyl lactate, and γ-butyrolactone was observed, compounded by a decrease in terpenes (mainly citronellol and geraniol). Albarín and Prieto Picudo wines had the highest concentration of amino acids. Overall, the levels of biogenic amines in all the sparkling wines studied were very low, which confirms the quality and safety of the samples.

González-Royo et al. [11] chose to assess the influence of sequential inoculation of some non-*Saccharomyces* (*Torulaspora delbrueckii* and *Metschnikowia pulcherrima*) and *Saccharomyces cerevisiae* yeasts on the composition and quality of the base wine. The data indicated an increase in the concentration of glycerol in the samples obtained with *Torulaspora delbrueckii* Biodiva™ (Lallemand Oenolog, France) as well as a reduction in volatile acidity. However, a positive effect on foaming capacity and foam persistence was noted. On the other hand, the strain *Metschnikowia pulcherrima*

Flavia® MP346 (Murphy & Son Limited, UK) caused both an increase in the persistence of the foam and a significant change in the aromatic profile by intensifying the smoky notes and the floral aroma.

Martí-Raga et al. [63] analyzed the effect of nitrogen content on the progress of the second fermentation. Three different strains were evaluated at two different fermentation temperatures (12°C and 16°C, respectively). The results showed that the nitrogen consumption during the second fermentation is very low, a fact that calls into question the common oenological practice of adding nitrogen to the base wine before fermentation, which only improves the fermentation kinetics in very old wines low in nitrogen (below 30 mg/L). As this particular study dealt primarily with the factors that affect the second fermentation, their effects on the sparkling wine, their impact on autolysis and the organoleptic characteristics of the end-product after the maturation period remain to be established.

Benucci [64] researched the effect of storage conditions after bottling on the color characteristics and sensory profile of some *rosé* sparkling wines. To that effect, the samples were bottled in Antique Green glass for 9 months following the disgorging stage. Different storage conditions were tested, sampled at different temperatures and different levels of lighting (30°C, in the dark; 5°C, in the dark; 5°C, under UV irradiation). The data indicated considerable variation in color intensity and hue compared to baseline values. In the case of the variants kept under UV irradiation, both an important reduction in color intensity (−22%) and a considerable increase in shade (+ 46%) were noted. Sparkling wines stored at 30°C, in the dark showed a 16% increase in color intensity and a 33% increase in hue. The variants kept at 5°C, in the dark exhibited an increase of only 4% in color intensity and 9% in hue. All samples presented significant differences in sensory profile. Thus, the first category of sparkling wines (30°C, in the dark) was associated with intense notes of burned material, while the last category of samples (5°C, UV irradiation) gave off unpleasant aromas, such as the smell of wet wool.

Caliari et al. [65] compared sparkling wines obtained from the Moscato Giallo variety produced by traditional methods, Charmat and Asti. As such, they proposed to analyze the volatile profile of the final samples. Sparkling wines produced by the traditional method had the highest concentration of volatile compounds (especially 2-phenylethanol, ethyl octanoate, linalool, and α-terpineol), while those produced by the Asti method showed the lowest concentrations. Principal component analysis confirmed the major bearing of production methods on the volatile composition of Moscato Giallo sparkling wines. These results point out that the method used to produce sparkling wines significantly influences the volatile composition of the final product.

6.1 Use of reverse osmosis

Sparkling wines entail the secondary fermentation of still wines. The quality and characteristics of the sparkling wine, the degree of stability, and the sensory properties are largely dependent on the physical–chemical composition of the grapes and the base wine, the production technology applied, the environmental conditions, etc. [39]. Currently, the alcoholic concentration of wines tends to be higher due to numerous variables, especially climate change [66]. At the same time, many consumers prefer low-alcohol products (9–13% vol. alc.) as a consequence of social aspects (traffic restrictions), but also of health problems associated with frequent consumption [67]. In recent years, several techniques have been tried out to produce wines with a low-alcohol concentration, either by using must with lower sugar concentrations, certain selected yeasts, or by the early interruption of alcoholic fermentation [68]. Also, in order to reach a predetermined alcohol concentration in wines, various practices, such as thermal or membrane-based processes, can be used. Thermal treatments often involve the degradation of some volatile compounds of interest [69]. Various membrane-based procedures are known to reduce the alcoholic concentration of wines while also contributing to preserving the initial sensory properties [68]. These procedures (nanofiltration, reverse osmosis) have a number of advantages, such as low energy consumption when working at low to moderate temperatures. Reverse osmosis involves passing the wine through a very fine filter to separate the alcohol. A certain amount of the alcohol is usually removed before combining all the elements (including the color and flavor compounds that have been filtered out) together again. It can also be used to reduce the amount of water in wine or must to concentrate the aroma compounds. The process can be used to correct some defects: reducing *Brett* flavor, volatile acidity, and smoke notes in a wine [70].

Carried out at low temperatures, reverse osmosis is successfully used to reduce the alcohol concentration since it generates a minor negative impact on the structure and composition of the wine. As such, it ensures the stability of the aromatic compound and the sensory characteristics [70]. As Pham et al. [71] point out, the application of reverse osmosis has a prominent effect on the quality characteristics of some red wines. Thus, besides the decrease in alcohol concentration and the loss of free sulfur dioxide, the reduction of some organic acids and total acidity were also noted, together with an increase in volatile acidity and pH. Moreover, significant effects were recorded concerning the color of the samples and astringency associated with a change in the content of phenolic compounds and anthocyanin, respectively. With regard to the volatile fraction, an important decrease in ethyl esters was obtained with the reduction in the alcohol content.

The impact of the reverse osmosis process is subject to the operating conditions applied. In this respect, Ivić et al. [72] studied the influence of variable pressure levels (2.5–5.5 MPa), with or without cooling, on some red wines. The retention of phenolic compounds depends on several factors such as the type of membrane used, the size of the pores, the polarity of the membrane and the compounds, the various interactions between the compounds and the membrane, etc. The results indicated that applying a higher pressure and cooling the retainer favor the retention of phenolic compounds and also trigger a stronger antioxidant activity. Changing the pressure did not alter the chromatic characteristics of the polymer significantly, while a high temperature generated by the absence of a cooling regime yielded a higher polymer color associated with the degradation of anthocyanin. These results were confirmed by Cotea et al. [39], who used the reverse osmosis process to obtain the base wine for the production of some sparkling wines in correlation with an appropriate cooling regime. A corresponding decrease in the alcoholic concentration of the base wine from 12.5 to 10.5% vol. alc. was recorded. The available data shows this procedure to be an effective alternative for reducing the alcohol concentration on account of its negligible influence on the physical–chemical properties of the base wines.

Research on using reverse osmosis to obtain wines with reduced alcohol concentration is limited. The causes could be the high cost of the devices and the reduced efficiency in terms of the quantity obtained in a certain time interval, correlated with energy consumption. Also, the complexity of this method requires knowledge from various domains for the efficient adjustment of the parameters.

7. Conclusions

Technological conditions directly affect the chemical composition and structure of sparkling wines, which calls for a well-structured work protocol in keeping with the intended purpose. The type of inoculated yeasts for the two fermentations is vital for defining the phenolic profile, the content of organic acids, metal ions, the volatile fraction, and the sensory characteristics of sparkling wines. Given the growing consumer trend favoring low-alcohol drinks, reverse osmosis is a worthwhile alternative technique for beverages with low alcohol content due to its minimal impact on the quality of the final product.

Author details

Camelia Elena Luchian, Elena Cristina Scutarașu, Lucia Cintia Colibaba, Mihai Cristian Focea and Valeriu Cotea*
Iași University of Life Sciences, Iași, Romania

*Address all correspondence to: vvcotea@yahoo.com

References

[1] Cotea V. Tehnologia vinurilor efervescente. Bucharest: Romanian Academy; 2005

[2] Muñoz-Redondo JM, Ruiz-Moreno MJ, PuertasB,Cantos-VillarE,Moreno-RojasJM. Multivariate optimization of headspace solid-phase microextraction coupled to gas chromatography-mass spectrometry for the analysis of terpenoids in sparkling wines. Talanta. 2020;**208**:120483

[3] Butnariu M. Biotechnology of flavored or special wines. In: Grumezescu A, Holban AM, editors. Biotechnological Progress and Beverage Consumption. Vol. 19. India: Woodhead Publishing; 2020. pp. 253-282

[4] Gougeon RD, Reinholdt M, Delmotte L, Miéhé-Brendle J, Chézeau JM, LeDred R, et al. Direct observation of polylysine side-chain interaction with smectites interlayer surfaces through 1H-27Al heteronuclear correlation NMR spectroscopy. Langmuir. 2002;**18**:3396

[5] Gougeon RD, Soulard M, Miéhé-Brendle J, Chézeau JM, LeDred R, Jeandet P, et al. Analysis of two bentonites of enological interest before and after commercial activation by solid Na_2CO_3. Journal of Agricultural and Food Chemistry. 2003;**51**:4096

[6] Gougeon RD, Soulard M, Reinholdt M, Miéhé-Brendle J, Chézeau JM, LeDred R, et al. Polypeptide adsorption onto a synthetic montmorillonite: A combined solid-state NMR, X-ray diffraction, thermal analysis and N_2 adsorption study. European Journal of Inorganic Chemistry. 2003;**2003**:1366

[7] DiGianvito P, Arfelli G, Suzzi G, Tofalo R. New trends in sparkling wine production: Yeast rational selection. In: Grumezescu AM, Holban AM, editors. Alcoholic Beverages. Cambridge, UK: Woodhead Publishing; 2019. pp. 347-386

[8] Garofalo C, Arena MP, Laddomada B, Cappello MS, Bleve G, Grieco F, et al. Starter cultures for sparkling wine. Fermentation. 2016;**2**:21

[9] Perpetuini G, Di Gianvito P, Arfelli G, Schirone M, Corsetti A, Tofalo R, et al. Biodiversity of autolytic ability in flocculent *Saccharomyces cerevisiae* strains suitable for traditional sparkling wine fermentation. Yeast. 2016;**33**(7):303-312. DOI: 10.1002/yea.3151.2016

[10] Capozzi V, Tufariello M, Berbegal C, Fragasso M, De Simone N, P, Venerito P, Bozzo F, Grieco F. Microbial resources and sparkling wine differentiation: State of arts. Fermentation. 2022;**8**:275. DOI: 10.3390/fermentation8060275

[11] González-Royo E, Pascual O, Kontoudakis N, Esteruelas M, Esteve-Zarzoso B, Mas A, et al. Oenological consequences of sequential inoculation with non-*Saccharomyces* yeasts (*Torulaspora delbrueckii* or *Metschnikowia pulcherrima*) and *Saccharomyces cerevisiae* in base wine for sparkling wine production. European Food Research and Technology. 2014;**2014**:999-1012. DOI: 10.1007/s00217-014-2404-8

[12] Palacios S, Vasserot Y, Maujean A. Evidence of sulfur products adsorption by yeast lees. American Journal of Enology and Viticulture. 1997;**1997**:48-525

[13] Vasserot Y, Steinmetz V, Jeandet P. Study of thiol consumption by yeast lees. International Journal of Molecular Microbiology. 2003;**83**:201

[14] Aguilera F, Peinado RA, Mill C, Ortega JM, Mauricio JC. Relationship between ethanol tolerance, Hþ-ATPase activity and the lipid composition of the plasma membrane in different wine yeast strains. International Journal of Food Microbiology. 2006;**110**:34e42

[15] Bauer EF, Pretorius IS. Yeast stress response and fermentation efficiency: How to survive the making of wine e a review. South African Journal of Enology and Viticulture. 2000;**21**:27-51

[16] Arneborg N, Hoy CE, Jorgensen OB. The effect of ethanol and specific growth rate on the lipid content and composition of Saccharomyces cerevisiae grown anaerobically in a chemostat. Yeast. 1995;**1**:953e959

[17] Borrull A, Lopez-Martinez G, Miro-Abella E, Salvado Z, Poblet M, Cordero-Otero R, et al. New insights into the physiological state of *Saccharomyces cerevisiae* during ethanol acclimation for producing sparkling wines. Food Microbiology. 2015;**54**:20-29. DOI: 10.1016/j.fm.2015.11.001

[18] Valero E, Millan C, Ortega JM. Influence of oxygen addition during growth phase on the biosynthesis of lipids in *Saccharomyces cerevisiae* (M, 30-9) in enological fermentations. Journal of Bioscience and Bioengineering. 2001;**92**:33-38

[19] Cebollero E, Gonzalez R. Induction of autophagy by second-fermentation yeasts during elaboration of sparkling wines. Applied and Environmental Microbiology. 2006;**72**:4121-4127

[20] Ribéreau-Gayon P, Dubourdieu D, Donèche B, Lonvaud A. Trattato di enologia i microbiologia del vino. Bologna, Italy: Edagricole; 2004

[21] Marchal R, Lallement A, Jeandet P, Establet G. Clarification of Muscat musts using wheat proteins and the flotation technique. Journal of Agricultural and Food Chemistry. 2003;**51**:2040

[22] Marchal R, Marchal-Delahaut L, Michels F, Parmentier M, Lallement A, Jeandet P. Wheat gluten used as clarifying agent of musts and white wines. American Journal of Enology and Viticulture. 2002;**53**:308

[23] Tracey RP, Britz TJ. The effect of amino acids on malolactic fermentation by *Leuconostoc oenos*. Journal of Applied Bacteriology. 1989;**67**:589

[24] Jeandet P, Vasserot Y, Liger-Blair G, Marchal R. Sparkling wine production. In: Concise Encyclopedia of Science and Technology of Wine Publisher. CRC Press; 2021. pp. 471-487

[25] Marchal R, Chaboche D, Marchal-Delahaut L, Gerland C, Gandon JP, Jeandet P. Detection and quantification of lysozyme in Champagne wines. Journal of Agricultural and Food Chemistry. 2000;**48**:3225

[26] Gnoinski GB, Schmidt SA, Close DC, Goemann K, Pinfold TL, Kerslake FL. Novel methods to manipulate autolysis in sparkling wine: Effects on yeast. Molecules. 2021;**26**(2):387. DOI: 10.3390/molecules26020387

[27] LaGatta B, Picariello G, Rutigliano M, Lopriore G, Petrella G, Rusco G, et al. Addition of lees from base wine in the production of Bombino sparkling wine. European Food Research and Technology. 2016;**242**(8):1307-1317. DOI: 10.1007/s00217-016-2634-z

[28] Raymond, Eder ML, Rosa AL. Non-conventional grape varieties and yeast starters for first and second fermentation in sparkling wine production using the traditional method. Fermentation.

2021;**7**(4):321. DOI: 10.3390/fermentation7040321

[29] Ivit NN, Kemp B. The impact of non-Saccharomyces yeast on traditional method sparkling wine. Fermentation. 2018;**4**:73

[30] Canonico L, Comitini F, Ciani M. *Torulaspora delbrueckii* for secondary fermentation in sparkling wine production. Food Microbiology. 2018;**74**:100-106. DOI: 10.1016/j.fm.2018.03.009

[31] Ivit NN, Loira I, Morata A, Benito S, Palomero F, Suarez-Lepe JA. Making natural sparkling wines with non-*Saccharomyces* yeasts. European Food Research and Technology. 2018;**244**:925-935. DOI: 10.1007/s00217-017-3015-y

[32] Alexandre H, Guilloux-Benatier M. Yeast autolysis in sparkling wine – A review. Australian Journal of Grape and wine research. 2006;**12**:119-127

[33] Molnar I, Oura E, Suomalainen H. Study of volatile substances produced during autolysis of Champagne yeasts. Acta Alimentaria. 1981;**10**:27-36

[34] Nunez YP, Carrascosa AV, Gonzalez R, Polo MC, Martinez-Rodriguez AJ. Effect of accelerated autolysis of yeast on the composition and foaming properties of sparkling wines elaborated by a *champenoise* method. Journal of Agricultural and Food Chemistry. 2005;**53**:7232-7237

[35] Martínez-Rodríguez AJ, Polo MC, Carrascosa AV. Structural and ultrastructural changes in yeast cells during autolysis in a model wine system and in sparkling wines. International Journal of Food Microbiology. 2001a;**71**:45-51

[36] Martínez-Rodríguez AJ, Carrascosa AV, Barcenilla JM, Pozo-Bayón MA, Polo MC. Autolytic capacity and foam analysis as additional criteria for the selection of yeast strains for sparkling wine production. Food Microbiology. 2001b;**18**:183-191

[37] International Organisation of Vine and Wine. Compendium of International Methods of Analysis of Vines and Musts. Dijon, France; 2023

[38] Hardy G. Les différentes methodes d'élaboration des vins effervescents. Revue des Oenologues. 2003;**107S**:9

[39] Cotea VV, Focea MC, Luchian CE, Colibaba LC, Scutarașu EC, Niculaua M, et al. Influence of different commercial yeasts on volatile fraction of sparkling wines. Food. 2021;**10**(2):247. DOI: 10.3390/foods10020247

[40] Vigentini I, Cardenas SB, Valdetara F, Faccincani M, Panont CA, Picozzi C, et al. Use of native yeast strains for in-bottle fermentation to face the uniformity in sparkling wine production. Frontiers in Microbiology. 2017;**8**:1225

[41] Tufariello M, Rizzuti A, Palombi L, Ragone R, Capozzi V, Gallo V, et al. Non-targeted metabolomics approach as a tool to evaluate the chemical profile of sparkling wines fermented with autochthonous yeast strains. Food Control. 2021;**126**:108099

[42] Tufariello M, Chiriatti MA, Grieco F, Perrotta C, Capone S, Rampino P, et al. Influence of autochthonous *Saccharomyces cerevisiae* strains on volatile profile of negroamaro wines. LWT-Food Science and Technology. 2014;**58**:35-48

[43] Alfonzo A, Francesca N, Matraxia M, Craparo V, Naselli V, Mercurio V, et al.

Diversity of *Saccharomyces cerevisiae* strains associated to racemes of grillo grape variety. FEMS Microbiology Letters. 2020;**367**:fnaa079

[44] Bozdogan A, Canbas A. Influence of yeast strain, immobilisation and ageing time on the changes of free amino acids and amino acids in peptides in bottle-fermented sparkling wines obtained from *Vitis vinifera* cv. Emir. International Journal of Food Science. 2011;**46**:1113-1121

[45] Medina-Trujillo L, González- Royo E, Sieczkowski N, Heras J, Canals JM, Zamora F. Effect of sequential inoculation (*Torulaspora delbrueckii*/ *Saccharomyces cerevisiae*) in the first fermentation on the foaming properties of sparkling wine. Bio Web of Conferences. 2016;**7**:02024. DOI: 10.1051/bioconf/20160702024

[46] Velázquez R, Zamora E, Álvarez ML, Ramírez M. Using *Torulaspora delbrueckii* killer yeasts in the elaboration of base wine and traditional sparkling wine. International Journal of Food Microbiology. 2019;**289**:134-144

[47] Loureiro V, Malfeito-Ferreira M. Spoilage yeasts in the wine industry. International Journal of Food Microbiology. 2003;**86**:23-50

[48] Pozo-Bayón MA, Martinez- Rodriguez A, Pueyo E, Moreno- Arribas MV. Chemical and biochemical features involved in sparkling wine production: From a traditional to an improved winemaking technology. Trends in Food Science & Technology. 2009;**20**:289-299. DOI: 10.1016/j.tifs.2009.03.011.53

[49] Cotea DV, Zănoagă C, Cotea VV. Tratat de oenochimie. București: Editura Academiei Române; 2009

[50] Jackson RS. Wine Science: Principles and Applications. Food Science and Technology. USA: Academic Press; 2008

[51] Waterhouse AL, Sacks GL, Jeffery DW. Understanding Wine Chemistry. UK: John Wiley & Sons; 2016

[52] Berger RG. Flavours and Fragrances: Chemistry, Bioprocessing and Sustainability. Berlin: Springer; 2007

[53] Bosch-Fuste J, Riu-Aumatell M, GuadayolJM, CalxachJ, Lopez-TamamesE, Buxaderas S. Volatile profiles of sparkling wines obtained by three extraction methods and gas chromatography mass spectrometry (GC-MS) analysis. Food Chemistry. 2007;**105**(1):428-435

[54] Ferreira RB, Monteiro S, Piçarra-Pereira MA, Tanganho MC, Loureiro VB, Teixeira AR. Characterization of the proteins from grapes and wines by immunological methods. American Journal of Enology and Viticulture. 2000;**51**:22-28

[55] Pons-Mercadé P, Giménez P, Gombau J, Vilomara G, Conde M, Cantos A, et al. Oxygen consumption rate of lees during sparkling wine (Cava) aging; influence of the aging time. Food Chemistry. 2020;**342**:128234. DOI: 10.1016/j.foodchem.2020.128238

[56] Martinez-Lapuente L, Guadalupe Z, Ayestaran B, Ortega-Heras M, Perez-Margarino S. Changes in polysaccharide composition during sparkling wine making and aging. Journal of Food Agricultural and Food Chemistry. 2013;**61**:12362-12373. DOI: 10.1021/jf403059p

[57] García MJ, Aleixandre JL, Álvarez I, Lizama V. Foam aptitude of Bobal variety in white sparkling wine elaboration and study of volatile compounds.

European Food Research Technology. 2009;**229**:133-139

[58] Hidalgo P, Pueyo E, Pozo-Bayón MA, Martinez-Rodriguez AJ, Martin-Alvarez P, Polo MC. Sensory and analytical study of rose sparkling wines manufactured by second fermentation in the bottle. Journal of Agricultural and Food Chemistry. 2004;**52**:6640-6645

[59] Mas A, Padilla B, Esteve-Zarzoso B, Beltran G, Reguant C, Bordons A. Taking advantage of natural biodiversity for wine making: The WILDWINE project. Agriculture and Agricultural Science Procedia. 2016;**8**:4-9

[60] Ruiz-Moreno MJ, Muñoz- Redonto JM, Cuevas FJ, Marrufo- Curtido A, Leon JM, Ramírez P, et al. The influence of pre-fermentative maceration and ageing factors on ester profile and marker determination of Pedro Ximenez sparkling wines. Food Chemistry. 2016;**230**:697-704. DOI: 10.1016/j.foodchem.2017.03.048

[61] Tofalo R, Perpetuini G, Rossetti AP, Gaggiotti S, Piva A, Olivastri L, et al. Impact of *Saccharomyces cerevisiae* and non-*Saccharomyces* yeasts to improve traditional sparkling wines production. Food Microbiology. 2022;**108**:104097. DOI: 10.1016/j.fm.2022.104097

[62] Pérez-Magariño S, Ortega-Heras M, Martínez-Lapuente L, Guadalupe Z, Ayestarán B. Multivariate analysis for the differentiation of sparkling wines elaborated from autochthonous Spanish grape varieties: Volatile compounds, amino acids and biogenic amines. European Food Research and Technology. 2013;**236**:827-841. DOI: 10.1007/s00217-013-1934-9

[63] Martí-Raga M, Peltier E, Mas A, Beltran G, Marullo P. Genetic causes of phenotypic adaptation to the second fermentation of sparkling wines in *Saccharomyces cerevisiae*. Genes. 2017;**7**:399-412

[64] Benucci I. Impact of post-bottling storage conditions on colour and sensory profile of a rosé sparkling wine. LWT – Food Science and Technology. 2019;**2019**. DOI: 10.1016/j.lwt.2019.108732

[65] Caliari V, Panceri CP, Rosier JP, Bordignon-Luiz MT. Effect of the traditional, Charmat and Asti method production on the volatile composition of Moscato Giallo sparkling wines. Lebensmittel-Wissenschaft & Technologie. 2015;**61**:393-400

[66] Jones JE, Kerslake FL, Close DC, Dambergs RG. Viticulture for sparkling wine production: A review. American Journal of Enology and Viticulture. 2014;**65**:407-416

[67] Saliba A, Ovington LA, Moran C. Consumer demand for low-alcohol wine in an Australian sample. International Journal of Wine Research. 2013;**2013**:1-8

[68] Pickering GJ, Heatherbell DA, Vanhanen LP, Barnes M. The effect of ethanol concentration on the tem-poral perception of viscosity and density in white wine. American Journal of Enology and Viticulture. 1998;**49**:30-318

[69] Takacs L, Vatai G, Korány K. Production of alcohol free wine by pervaporation. Journal of Food Engineering. 2007;**78**:118-125

[70] Saha B, Torley P, Blackmann JW, Scmidtke LM. Review of processing technology to reduce alcohol levels in wines. In: Proceedings of the 1st International Symposium Oenoviti International. Bordeaux, France; 2013

[71] Pham DT, Stockdale VJ, Wollan D, Jeffery DW, Wilkinson KL. Compositional consequences of partial dealcoholisation of red wine by reverse osmosis-evaporative perstraction. Molecules. 2019;**24**:1404. DOI: 10.3390/molecules24071404

[72] Ivíc I, Kopjar M, Buljeta I, Pichler D, Mesíc J, Pichler A. Influence of reverse osmosis process in different operating conditions on phenolic profile and antioxidant activity of conventional and ecological Cabernet Sauvignon red wine. Membranes. 2022;**12**:76. DOI: 10.3390/membranes12010076

Chapter 6

Innovation and Entrepreneurial Practices in the Wine Sector: Constant Experimentation and Playfulness

Christian Poulsen and Mette Mønsted

Abstract

Winemaking is a complex process of viticulture followed by a fermentation process with subsequent control of taste development. There are some stages in wine production where innovation practices may be applied: Grape variety, fermentation techniques, skin contact, whether to apply organic production and the collaboration with an oenologist or not are all focus points of innovation efforts. These efforts would have to cohabit in a wine sector where tradition and experience are highly valued. The authors have been studying how experimentation and practices of innovation take place in the wineries at the individual level and at different stages of wine production. The innovation processes are not radical innovation but tend to be entrepreneurial practices of experimentation. The article takes a first glance at how practice theory can be implemented in the study of wine and entrepreneurship. It investigates the relationship between winemaking practice and playfulness and experimentation to make a wine that stands out. The chapter is based on qualitative interviews in New South Wales (Australia), Var (France) and Ribera del Duero and Toro (Spain).

Keywords: entrepreneurial practices, experimentation, playfulness, winemaking, individual practice of innovation

1. Introduction: What is innovation in the wine sector?

How do you discuss innovations and unique profiling for wineries in a sector which, even if they do have new technology and education, in so many countries have been leaning on tradition, practice, and old skills in the family? In the empirical data we have collected, innovation is not radical, which confirms a tendency in the industry [1] but tends to be based on entrepreneurial behaviour and experimentation in practice.

The many production processes of very different nature, and the time perspective of maturing, make the wine field a very interesting field for the study of innovation and entrepreneurial action. The way wine is talked about by experts and connoisseurs is part of the status and special features of the wine sector. No other agricultural sector has this glamour. Making wine includes both pure production aspects as artistic aspects, like performativity.

In this field study we have investigated at the micro level, how the individual growers and wineries tackle innovation challenges, new practices and create a special profile of the wine in relation to others in the region. We have sought to explore how the entrepreneurial actions are revealed in their practice. The focus of our study is how they describe and argue for the experimentation and playfulness of the new methods. The uncertainty of the outcome of experiments is high and individualised, and in this way, it resembles more drastic innovation processes, even if they have to re-connect with traditional *savoir faire*.

New technologies for testing in the wineries' own laboratories have provided explicit knowledge to supplement the constant evaluation of implicit or tacit knowledge by the winemakers. All claim that tasting is necessary, as the tacit knowledge of wine production is a fundamental part of wine production. The blending is a "magical moment" and impossible to explain explicitly. The wine sector thus is an interesting case of how the explicit and tacit knowledge are complementary in the innovative practice. Experiments are developed constantly, but the results have to wait at least a year to be assessed.

There has been a search for a strategic approach to legitimising innovation [2], and this article aims at filling some of this void. The field of literature on innovation has less focus on practice on the micro level, whereas the research leading to this paper will be directed at the micro-level entrepreneurial practices. We are focusing on how constant experimentation is contributing to new profiles and features of the wine, and a possibility to distinguish the wine from the neighbouring wineries' wine. The Research Question is: How does constant experimentation contribute to the practice of innovation in wineries?

We have conducted a series of case studies in Australia, France and Spain to research their perspective, strategy and practice on innovation. Our research indicates that practice theory shows promising results in understanding innovation, entrepreneurial practices, experiments and uncertainty management, in the wine sector.

The chapter presents our analysis of the case studies and are presented thematically according to the themes that appeared in our interviews. The themes reflect the different stages in wine production and are presented in a sequence to mirror the winemaking process.

In Section 2 we present the tentative practice framework that is being used.

Section 3 presents the data and the methods are described.

Then we follow the process of winemaking, as Section 4 covers the innovation in viticulture. The subsections look at how grape variety, managing sunlight, irrigation and water, organic or green management, harvesting, pressing and destemming technologies influence innovation efforts.

Section 5 is devoted to innovation in the winemaking or cellar work. The subsections are fermentation, lighter tasting wines with lower alcohol percentage and blending.

Section 6 is discussing the use of professional winemakers.

Section 7 takes up the analysis and discussion of innovation and entrepreneurial practices, with subsections on experiments and playfulness under condition of uncertainty, and technology and process innovation.

In Section 8 we present the conclusion.

2. A tentative practice framework

Innovation in the wine sector is not easily identified, and we have chosen to not only rely on innovation literature, but also on entrepreneurial practice, improvisation and experimentation in a knowledge management perspective, as we are looking at the practice of viticulture and winemaking.

Most of the literature on innovation in the wine sector is concerned with clusters or systems of innovation, i.e., a regional level of analysis [3–6], and in relation to wine districts. Thus, Australian wineries are better in organising and have better R&D support facilities, than other New World countries like New Zealand, South Africa and California [7, 8]. Yet others emphasise, "that innovation implementation is a firm-level activity, which should be affected by individual-level and country-level culture" [9]. A sub-field of the regional perspective leans on building networks and use of knowledge brokers in the region [1, 10]. The use of official brokers at the regional level was rather absent in our data, whereas the role of consultant winemakers seems to play an important role. The regional perspective literature tends to stay at the macro level, and does not allow for recognising knowledge as embodied and local, thus focusing on the individual practice and the process of entrepreneurial practices [11].

Theoretically, we define innovations as new methods, new processes and new ways of organising [2]. "The innovation process itself, which involves design, development, implementation and mastery of novel ideas, takes place within an institutional and social context with people engaged in the process bound by repeated transactions over time" ([12], p. 315). Knowledge in the wineries seems to become new knowledge in parallel to the knowledge conversion process [13].

Often the discussion of innovation is around the binary of science versus art or whether wine is made in the vineyard or in the winery, and it does not appear to be in either or [14] but in both ([15], p. 243). The few inventions in the history of winemaking "have actually enforced the strength of the binary," as McIntyre refers "the characters imparted by terroir are expressed more fully as a result of temperature-controlled fermentation." ([15], p. 244). Thus, while temperature-controlled fermentation has revolutionised the vinification process, it has strengthened the role of traditional viticulture. Tradition is not a mere condition of institutional inertia but a praised virtue in the wine sector, and a part of the tacit knowledge in the sector.

The basic idea of terroir, which is certified in the "Appellation d'Origine Controlée (AOC [1])," is the importance of preserving the sensorial expressions such as taste, smell and colour of the terroir. "Where a wine, or more specifically the grapes originate from is important" ([8], p. 510), thus creating path dependency [2]. Other articles on innovation in the wine sector stress the lack of radical innovation [16], and the incremental innovation or small changes, such as Farias and Tatsch [1] on the Brazilian case, Dressler [4] on German winemaking, Lenzi [17] on local governance of innovation in an Italian case or Castellano & Khelladi [2] on the use of open innovation in the French wine sector. Innovation in the wine sector is conditioned by the high esteem that the traditional way of doing things in the local area or in the family has. The path dependency on tradition is voluntary to a degree we do not see in other areas of food production; the high membership rate of the Denomination of Origin, such as "Denominaciones de Origen (DO)" and "Appelation d' Origine Controlé

[1] Denominación de Origen (DO) and Denominación de Origen Cualificada (DOC) in Spain. For ease of reading, we will use AOC throughout the text.

(AOC)"[2] being the prime indicator; "yet innovation is not exclusive to the old world," as Lukacs [14] points out.

The focus on cumulative innovation and entrepreneurial behaviour lends itself to a practice theory perspective on innovation. The practice perspective offers a possibility to look at the "how" of human actions by breaking social activities into several elements [18]. A practice consists of bodily activities, mental activities, tools, background knowledge, states of emotion and motivational knowledge ([18], p. 813). The focus on everyday practice allows researchers to look into how wine workers do the management of muddling through, rather than focus on contrasted success stories [11]. This focus also allows us to connect innovation to knowing, as "know how" in contrast to "know that" is linked to performative, sticky knowledge [19]. As Gherardi has emphasised knowing in practice can be articulated as a "creative entanglement of knowing and doing" [20]. By investigating the practices of innovation in wineries, our aim is to study how tacit and explicit knowledge serve as the ground where epistemic work is done between knowledge and knowing, e.g., how innovation evolves [21].

Experimentation in the wine sector is beyond the notion of individual strokes of genius and aligned with the theory of communities of practice; we see collaboration as central to innovation [19, 20]. Communities of Practice consist of people, who share practice, "They are likely to have communal knowhow from that practice" ([19], p. 204). The experimentation could be interpreted as an entrepreneurial decision making [22]. The craft of vine cultivation and winemaking as doing and experiences cover well the perception of learning by doing, and how "people solve practical problems that have emerged as a result of specific and unplanned circumstance/entrepreneurial opportunities" ([22], p. 12). Improvisation embraces creative uncertainty within structured regimes [23]. However, it is important to emphasise that there is much preparation and study behind effective improvisation [24]. It "relies on rules and routines that are pre-established and rehearsed" ([25], p. 203), and knowledge as investigated here "is a capacity to act within a situation" [26]. Improvisation is the search for "making do" rather than "letting go," as it is a search for novelty and usefulness ([26], p. 205). This perspective is very close to Sarasvathy's theory and concepts of entrepreneuring as effectuation [27].

3. Data and methods

The study is based on 11 face-to-face interviews with wineries and three with experts in New South Wales (Hunter Valley, Orange and Riverina), seven wineries in Var in the South of France, and five wineries and one expert from the Ribera del Duero demarcated region in the Northwest of Spain. Most interviewees are owners or oenologist or both as they are found to be closest to the innovation process [17]. Most of the interviewees represent the middle-range quality wine of small producers, but a few are very large in Australia and Spain.

The interviews were all conducted *in situ*, and observations in the vineyards and cellars were noted and served as a useful contextual supplement to the interviews. In nearly all the interviews, the conversations were conducted in the main building or cellar and included field visits, thus enabling the interviewees to show tools or

[2] In the following we will refer to Appelation d'Origine Contrôlée and Denominación de Origen with their abreviations AOC and DO.

knowledge assets they work with in order to discuss the different methods for differentiating the wine production.

We have employed respondent-driven sampling [28] combined with theoretical sampling as the version of grounded theory methods presented by Glaser [29–31]. We used referrals from experts and from other wineries to find wineries with a distinct profile in order to avoid the standard table-wine producers in the area, who would have dominated, if we had a representative sampling [32]. We have organised our sampling to look for new patterns that have emerged in the field. For example, in Domaine Tempier in the Bandol region in the South of France, a sommelier from one of the local restaurants who purchases wine in the tasting room, and in the discussion of who is innovative and extraordinary in the Bandol area, both he and the owner recommended the Domaine la Bégude winery. Personal network helped us gather informants in three countries, but once arrived in each country the described respondent-driven sampling was conducted.

Our interview guide was thus structured in few themes void of a sequential order. Our point of departure was a single pre-conceived idea of where differentiation, innovation and extraordinary methods could be found. That made up for a broad interview theme concerning the innovation of the wine producer with the subthemes of innovation in the vineyard, the cellar and marketing. The design would allow the data to produce analytical categories and could generate themes from the interviewees' references and discourses as they were generated most frequently. We would look for reports from "war stories" to access the accounts for the creation of knowledge within the wineries [21].

In the course of gathering our data, we quickly observed "What the wine producers mentioned as efforts of differentiation and innovative efforts." However, our focus increasingly shifted to "How the wineries addressed these efforts." This theme concerns their experimentation and entrepreneurial actions as innovation practices.

Qualitative interviews were used in order to obtain more substance in how the wineries profile themselves. This methodology was deemed essential [33], as a series of qualitative interviews would allow the researchers to investigate across wineries in different contexts, with a sensible economic expenditure on field research, and with naturalistic records that could show the interaction between researcher and research object.

Interviews were conducted in English, French or Spanish, where the choice of interview language depended on how the interviewees felt at ease. Interviews of about 1-2 hours are transcribed and quotes and extracts from these are used in the analysis. We also include a list of main characteristics (in the appendix). In the interviewing, we have probed into how the entrepreneurs acted to differentiate themselves from the others, and how they experimented and did something new, as the concept of innovation was not easily understood.

We emphasise the effort to do something different, and experiment with new grapes and methods at the individual level. Probing was done in the interviews to invite the interviewee to stay with the details of innovation practice. The probing on specifics of concrete situations in everyday life at the vineyards yields importance to phronesis, practical knowledge [11]. Our understanding of practice is thus devoted to theory of action in that we do not focus narrowly on the intentionality of action, but rather on how patterns of practice are enacted, talked about and produced.

Wine marketing is an important aspect of the wine industry, and part of the stories of great wines, castles and wine-experts. However, the viticulturists, winemakers and owners in our sample were actively deselecting this issue in our interviews, and marketing issues will be limited to customer influence on light wine types and organic wines. It does, however, have effects of marketisation as qualculative behaviour [20, 34] [The term qualculative is by the author of the reference]. News and stories of marketing and markets in this sense play a role in how the actors calculate prices of their wine and calibrate the quality of the product [34].

The regional aspects and local restrictions tied to AOC and DO are analysed in another article. Basically, here we refer to the kind of restrictions tied to the regional quality assessments (see [35]).

4. Innovation in viticulture

4.1 Types of grape

When asked about, what makes the wineries special compared to neighbouring wine producers, the winegrowers and owners spontaneously start talking about the area and the types of grape they cultivate. The vineyard is central in their search for expressing their differentiation. The type of grape is very important, and they experiment with new types, such as Mourvèdre in the Var, outside the Bandol and coastal area (Christine Mylene in Domaine des Selves, Var) while Mourvèdre is a grape taken for granted and compulsory in the Bandol wines, it is not evident for the inland areas.

Gerald Naef in Patina Wines in Orange experimented with his white wines (Sauvignon blanc) to make them more like the French Poilly-Fumé with the "white smoke," rather than the New Zealand lighter and fruitier wines. Secondly, he is elaborating wine with the Riesling and Sauvignon blanc types for a better fit with the high altitude and cool area. Many of the Orange county wine producers emphasised the cool high mountain climate that resembled some of the French wine areas. Philip Shaw says he went to this area, because the climate is "very similar to Burgundy." The two examples show experimentation and management of climatic conditions.

Terry Dolle used the high cool climate to make his icewine based on Viognier, which is not planted by many in Australia, but he thought "it should be fresh and reflecting the climate, so I made it an icewine," and James Sweetapple tries the Gewürtztraminer and Riesling, both in the high areas of Orange.

In areas of AOC, there are a number of restrictions on what grapes can be grown, such as the requirement for Mourvèdre in Bandol, and Tempranillo (Tinta del país) in Ribera del Duero. But even here they must assess what other types of grapes they would like to blend.

Some interesting examples of developing old types of vine are found in DO Rueda, Northern Spain. The old vines had been cleared from 1989 with EU subsidies to produce less table wine and more quality wine. Some of the old vine types with very deep roots in sandy soils have survived, and recovered, and are now developed and cultivated in order to get a special and good wine, and it is cloned to get the taste and features of these grapes.

Antonio Arrévalo is certifying that this older clone of Verdejo growing in sandy soils with very long roots and the ability to have grapes lying on the sand directly create a flavour, which is more concentrated and therefore something special. Just like neighbouring Javier Sanz he is going through the trouble of separating and cloning rather than using a commercial Verdejo. The two Rueda grape growers are going back to their roots, literally, to get answers to new challenges.

The choice of grapes is in France involving some innovative entrepreneurial action on older vine:

> *Laugier: Viogner has been over grafted and planted from the beginning, when it wasn't known at all, Viognier, the grape; and it was quite smart, well, or risky to go this way, when it was not already very known, very successful, but we were, in the beginning the first ones with this grape.*
>
> *Interviewer: OK. So others would have Rolle? In this area, Vermentino.*
>
> *Laugier: At that time none, no one at all, now everybody have a few rows, to be able to make a few bottles, because it was quite efficient this grape.*

New types of grapes are sometimes introduced by overgrafting of existing vine, as was the case in Triennes and Jas D'Esclan, when the existing old vine did not provide good quality, but only quantity, and they wanted an organic production.

Radical choices of grapes, however, do have consequences for wineries. Triennes, e.g., is not allowed to use the AOC Var for their Viognier wine, as this grape is not allowed in the AOC.

4.2 Managing sunlight

The pruning of vines and the desuckering for premium wines are very intensive handwork, but mainly left to casual labour.

In the very intensive cultivation of Patina Wines (Orange), Naef is talking about "managing sunlight":

> *Gerald Naef: I definitely believe that the wine is made in the vineyard more so than in the winery. It's about managing sunlight, I suppose, more than anything else and just the amount of sunlight that you get on the fruit. It varies from red grapes to white grapes and especially the season. .... In a very wet year, you get a lot of leaf cover and you've got to look at removing some of those leaves. In a dry year, you're battling to get enough leaf cover and you end up getting some sunburn. The sunburn, I've learned from experience that sunburn gives bitter characters to a wine and if you don't get a little bit of sunlight onto your fruit, something they called 'dappled' sunlight to... What you really want is one layer of leaf over your white grapes, your red grapes not necessarily; your red grapes maybe need a little bit more sunlight.*
>
> *You kind of manage that so you get that "dappled light" on your white grapes, and I think probably on your red grapes as well, it's just that you are going to see more sun coming through onto your red grapes and less coming through onto your white grapes, if you manage it correctly. I think it's not so much being innovative; it's more just being a little bit pedantic and being small enough to be able to do that sort of thing. (Naef, Orange).*

This is a very intensive cultivation, a practice which most of the larger wine producers are not doing. The Naef quote describes very well the entrepreneurial practices he is taking to manage the sunlight. In a continuous effort of experimentation with leaf cover he is building knowledge of how to avoid bitter tones in his green grapes. He insists on managing his vineyard by being pedantic, thus holding his hand over, as the Italian term manegiare, the root of managing would mean, the nurtured grapes. The quote shows how knowing in practice has technology and doing intertwined. Performativity runs in parallel with scientific behaviour in the innovation practice.

All producers are concerned about the hot years and the impact on grapes, and the level of sugar versus taste:

> *Tyrrell's: we test [grapes] regularly in the vineyard, and we are testing pH, acid, and sugar, of which pH is the most important. As a general rule, during vintage, 4 o'clock, the winemakers, my son Chris and I, and the vineyard manager, go to the testing area. Any juice that's been tested that day has been kept in a bottle. So we'll look at the results and we'll taste the juice, and the two most important things are the pH and the taste of the juice. If it's red – let us say most of your answer is in your eyes. If you have got Shiraz and there's any greenness at all in the colour of that juice sample, the wine's not right, and it just does not matter what the chemistry test says, it's not right.*
>
> *(Tyrrell's, Hunter Valley)*

The new technologies of lab testing methods for the quality of grapes are in all cases supplemented with the manual, visual and tasting as traditional methods by the expert leaning on his expertise and tacit knowledge. Knowledge assets, such as sugar tests, do not constitute knowledge by itself. The winemaker will add his or her "personal knowing" to translate a given measured sugar level to the alcohol level in the finished wine [36]. The question is viewed similarly in Var, France. In Triennes, the winemaker also checked the grapes regularly, allowing himself to supplement automated testing with personal tasting.

The management of humidity and the management of sunlight are a series of tricks of the trade that the viticulturist can perform to manipulate with the natural conditions that are given to everyone in the area. The innovation can come from new ideas or as Antonio Arrévalo describes it from a technique that is developed in the past, but is re-used in a new setting. In terms of innovation this is interesting, as it shows the new use of knowledge and value of combined knowledge for special features of the wine in this area, and entrepreneurial perception of new possibilities.

4.3 Irrigation and water: managing water with technology

In many of the areas irrigation is a serious discussion, whether or not to irrigate, and in some areas like the hot Riverina in Australia, the river provides the possibility to get irrigation, which is a necessity in this hot area. But in many of the small wineries in more traditional vineyards, they do not irrigate and actually have lots of arguments against irrigation, as they have to get vines with very deep roots, and a less "diluted" flavour. Irrigation is related both to the types of vine, the conditions of the soil, access to water, regulation and climate. Some of the efforts to create innovations are tied to the water handling, both how to irrigate and in order to save water, rinse water and recycle water.

In Australia there is a licence to secure a balanced system for those producing more than 30,000 tonnes. This is to protect water resources in the river and the soil from pollution. De Bortoli in Riverina who is crushing 65,000 tonnes has been especially active and had a large research project paid by the family and the Government to create green technology in the production. Lindsay Gulliver is health safety and environmental manager and he runs many projects on wastewater, but is basically looking into different ways of saving energy and cleaning water. The wastewater problem was serious and had "odour footprints of something like 15 km." He works on this both with ammonia to bring the pH up, and with a new low-energy systems of aeration.

The latest project is having 10 mill $ funding from the Government to create different forms of clean/green technology. This is both "smart-pumps" that can work at night, filtration of water, low energy bottle line and bottle warmer system to help clean bottles and make labels stay on the bottles.

All of the large wine-factories above 30,000 tonnes have to work on the wastewater and environmental issues as these are part of a Governmental contract.[3]

Both the Matarromera group and Bodegas Fariña in Spain are controlling drops of water and feeding water in trenches digged along the rows of vine.

The decision when and how to irrigate can thus be supported by root electronic control, but the technological development of measuring the humidity and irrigation is also an important control variable for farming as such. Rosa Zarza is an oenologist who works in several wineries and has similar experiences from Toro:

> *Zarza: It has a double face. Now I can see on the sensor under the leaf that the humidity under the leaf is raising, and this is because the producer has treated the vines and the farmer does not tell me anything.*
>
> *I: Aha, so you can control it from your home?*
>
> ....
>
> *Zarza: Now I can go to Denmark and if the ground sensor says it is not raining and the weather forecast says sun, and still the leaf sensor says humid, then I know something is not as it should be. (Rosa Zarza, Oenologist, North of Spain).*

The new methods of controlling humidity makes long distance control much easier, and allows for involvement of oenologists on many separated fields.

Yet farmers in other areas such as Bandol and Rueda are bound by restrictions in the AOC (DO) of the area so they do not irrigate. The wineries, we have interviewed in these two areas, however do not see it as a problem, but a quality of the terroir.

The permissive legislation in Tierra de Castilla y León and Australia allows vintners to apply technology to improve the management of water. The application of technology relies on existing models of thought; that there does exist an optimum level of humidity by the leaves and root of the vine.

4.4 Organic or green management

In many ways we could ask, what is the innovation challenge in organic production? The demand is coming from the market, but it is a field, where many are making entrepreneurial and innovative efforts. We have met several examples of organic production and especially nearly organic production, as a kind of "reasonable farming," which in France is regulated by a contract with the Ministry of Agriculture to use less chemicals for insecticide, herbicide and fertiliser. Only a few of the organic wineries actually sell their wine under the organic label, as they claim it does not give any real value in the market. In relation to innovation, this is an interesting aspect, as the ideology and the effort are to avoid the many chemical additives in the different stages of the production of wine. However, all informants discuss how difficult this is,

[3] We do not, however, have information on the other large wineries, as we have interviewed people of a different profile.

and for exports, the demands for certification in the USA make it nearly impossible to fulfil these criteria. The need for innovation and new methods clearly exist but the methods so far do not meet the demands.

The concern of limiting chemicals is very clear in all the medium-sized and smaller wineries with quality wine. Some were explicit about their efforts to make the production organic.

Philip Shaw in Orange has been on a 24-year journey towards organic production, but sees the organic labels on Organic farming as unattractive as it allows the use of copper and sulphur on the field.

In another Orange county winery James Sweetapple is into holistic farming, which is an overall ideology of organic production and balance in nature. He distinguishes between chemicals that he labels "pesticides" and "natural elements." He has sheep in the area to eat the grass and produce manure, and he worked to put organic matters into the soil to soften it as well as compost tea and a special soft rock phosphate. All the concerns were to create balance in nature. These practices form part of his strategy to manage the question on whether to be an organic winery or not.

Both Domaine des Selves, Val d'Iris and Chateau Roselline in Var are not doing organic, but have a "terra Vitis" contract with the Ministry of Agriculture on reasonable farming. Chateau Roseline clearly indicates that the climate is important for doing this:

> *We are very lucky because we have the area with a lot of Mistral (wind), so after the rain, very often the Mistral blows the plant dry. And if we compare with a farmer in Bordeaux for example, we have to treat, but we are doing half of the treatments because of the Mistral. (Chapelle, Var).*

Domaine de Triennes is doing organic production: The white and red wines are organic, and this is indicated at the homepage. Mr. Laugier admits that the innovation strategy is to place the Rosé with bought grapes from a larger area as a cash cow. And these are not organic. It brings a lot of revenue to the winery, and it allows Laugier and his colleagues to experiment with the other wines they commercialise restricted to own grapes.

In Domaine Tempier:

> *Rougeot-Peyraud: ..It is the soil which is the important, and the Spanish Mourvèdre. The soil is important, but it is not the same and not exposed in the same way. The soil is clay and chalk (limestone)...Here are different flowers, and insects also, but some flowers and draining help. It is natural. Natural fertiliser. (Rougeot-Peyraud, Bandol).*

The strong identity as being viticulturists who serve as a starting point for their innovation strategy [35] is in this quote aligned with a position close to reasonable agriculture.

Only two of the wineries in Var have organic produce, and sell it as organic, yet Jas D'Esclan does not see it as a change nor as an innovation as they continue the old tradition.

> *Jas D'Ésclan it is organic since 1939. It never changed .... when we bought the estate in 2004 it was like that. (de Wulf, Var).*

Tradition seems to weigh heavier than organic argumentation. In Bandol they are more challenged:

> *Tamayan: We are organic since 2009, then certified. It is a lot of work with soil and the vine. This year is very complicated as there is a lot of wind, and it dries the soil. (Tamayan, Var).*

Yet they could not irrigate to solve this problem as they are not allowed by the AOC Bandol, who put restrictions on types of grape, fertilisers and irrigation.

The conditions for organic or reasonable farming in Var in the South of France are good, as the soil, the climate and the wind all support the relative dry leaves, and it appears as if the insect attacks and usual diseases are few.

Epifanio Rivera in Ribera del Duero puts it this way "I am ecologist as of philosophy." When asked whether he is certified organic he nevertheless puts the foot down "no because I am not denying modernity in agriculture." Epifanio puts a lot on emphasis in that he is treating the soil with as few products as possible to work with the field in as natural a way as possible. However, he is keeping the opportunity to use a chemical directly on the weeds on certain extraordinary occasions.

> *Right now we are working with vegetal covering although it is more difficult in the area, because of the lack of rain gauge and because of the rain. (Epifanio, Ribera del Duero).*

Antonio Arrévalo in Matapozuelos is belonging to the viticulturist of Rueda that targets for the higher price range of white wines; Arrévalo does strengthen that they use very small amounts of pesticides. He would use sulphur twice or three times in a production, not more than that: "That is biodynamic, but listen, I am not looking at the moon, nor the stars…"(Arrévalo, Rueda), and he is not certified organic.

The consensus on doing as little harm as possible to the soil, and allowing for some chemistry, although self-labelled as "ecologist as of philosophy" would however both in Arrévalo and Epifanios cases fit in to what the French would speak about as "reasonable farming."

It is a theme where innovations and experimentation are needed, and most of the wineries make experiments and many entrepreneurial efforts to reduce chemicals and come closer to organic agricultural production. None of the respondents finds it easy. When it is such a challenge, it has been a surprise that the wineries on the one hand are so concerned about ecological or nearly ecological farming and a practice of limiting chemicals, but still few of them go into certification or labelling on bottles. The marketing value apparently does not play an important role for this behaviour. All are concerned about the chemicals, but they do not argue that the taste of organic wine is better. The small steps towards organic wine production are along with other innovation efforts part of an experimentation to make a job "well done" or "comme il faut" as Gherardi and Perrotta find in their study of craftsmen [37].

4.5 Harvesting: managing technology and machines

The innovation has mostly been with better machines for harvesting, i.e., more adoption of new technology, than their own innovations. Most are stressing how important it is to harvest with machines, also because this can be done by night, which is better for the grapes. The price difference is quite substantial. Tyrrell's says $775 per tonne to hand-pick and $45 per tonne to machine pick.

The harvesting has changed in most places and for most vintners' grapes will be machine harvested. But there are regional AOC demands and certain high-profiled vines, or old vines, which are still harvested by hand, as machines destroy old vine

This appears to be seen as a quality stamp, though again only advertised on the homepage, and not on the bottles. So, in the sense of Drucker [38], of taking the novelty to the market, it is difficult to label it as an innovation. There are other benefits from machine harvesting, as this may be done by night, and therefore protect the grapes from the heat after picking.

4.6 Pressing and destemming technologies: Process innovation

A number of the Australian wine producers had an education and experience in machinery. They discussed the different machines and how they have been improved. Some of the older types of presses are used for the premium wines, even if they do not get all of the juices out. These old machines are gentler to the grapes. It seems to be one of the fields, where technological innovations are coming in, but this is mainly adoption of new technology rather than innovations.

An example of solving the harvesting and destemming problems is presented by Manuel Perez (Bodegas Fariña), as they take the machinery to the field:

> *Manuel Perez: Yes. We de-stem, we crush the grapes and we pump it into a tank in the field. The tank is carried by a truck. This tank is without oxygen, we introduce nitrogen or carbon, normally nitrogen to avoid the oxygen in there. ....From there we take it to the winery. In the winery we just use a pump and introduce it in the tank to start the fermentation. If we want to add yeasts or not, with the reds we do not use it. (Perez, Ribera del Duero).*

Destemming in the field is an example of process innovation that builds on the management of time and temperature. The management of temperature in the fermentation process was the major technological innovation of the 1980s, and it was done by controlling the temperature when the juice was in the metallic tanks. By destemming in the field, the temperature control is expanded to go beyond the cellar and can, in this sense, be seen as a radical process innovation.

There are different practices, where some are clearly related to technology and knowledge of machinery, and others are tied to the process of organisation of the production. Specifically in the wine sector, there are very different levels of technology improvement and innovation. It is a range from large government-funded water projects (De Bortoli), and exploitation of residuals as skins (Matarromera) to buying new machinery at smaller wineries. Innovation in wine production can be seen as a process where experimentation and formal knowledge are mutually interdependent and intertangled [20]. At the individual level in smaller wineries, this is also a part of the improvisation [22], where the entrepreneurs are "reworking precomposed material and designs in relation to unanticipated ideas conceived, shaped, and transformed" ([24], p. 544). An example of such Improvisation could be the movement of destemming equipment to the field as Perez outlined above.

5. Innovation in winemaking: happy accidents

Some of the wineries are involved in different kinds of innovations or experimentation in the winemaking process. The main new fermentation technology is old by now and found everywhere with the temperature adjusting steel tanks. However, other aspects call for experiments on fermentation, blending and ageing.

These are much more difficult to get a clear answer from the respondents, as it is much more a discussion of who are involved. It is a field where experiments take place, but they rarely define these issues as new methods.

5.1 Fermentation

Controlled fermentation seems to be pretty standardised now, e.g., Lukacs sees temperature management *"as the second most important technological innovation in the entire history of wine, following only the invention of securely sealed glass bottle"* ([14], p. 227). Temperature control is used during fermentation but also afterwards to stabilise the wine.

In a market of fresh wines based on Verdejo using commercial yeast, Javier Sanz develops luxury wine based on maceration with long skin contact. For the avant-garde of Rueda, like the two old family firms, Javier Sanz and Antonio Arrévalo, it has become obvious that the next step in their innovation strategy is to develop wines that aim for the market above 10 euro:

> *Antonio Arrévalo:... If we talk about prices, if we talk of the products that are in the market now, we can find a verdejo wine at 1-1,50 euros. ... you do not take care of the integral palate, nor the acidity which for me is very important.*
>
> *Interviewer: By all means, the nose, no?*
>
> *Arrévalo: Exactly, exactly. Well they have gone for the nose. They have been after the special aromas and have used banana, passion fruit, apples and they have forgot about the primary aroma, that is the terrain, the minerals, forgot the mouth most importantly, that a wine is long, that it is powerful, that it has a good degree (of alcohol), and a good balance between degree and acidity. Well like that ... (Arrévalo, Rueda).*

Even though we have observed steel tanks with temperature control everywhere, even with the smallest wine producer, not all wine is treated in these steel tanks. Westend Estates Wines, which is a pretty large winery, uses 60-year-old open concrete fermenters, which are used today for premium red wines.

Also in Chateau Roseline and Triennes they use concrete tanks for some table wine, even if the temperature control is not as good as in the steel tanks, but there are other benefits, as they can stir the solid part of grapes in the liquid (pigeage).

Also the time to ferment with the skin is used for experiments, and a specially strong and concentrated wine:

> *Dor: Le Grand Foudre 2012 is based on Merlot, experiment of fermenting in a large wooden tank for 3 weeks, and then clean the mass and skin and back to the tank again. It is very concentrated, and I only make this every 2nd year. (Dor, Var).*

Both French cases are based on experiments with skin contact and pigeage. This is a very clear effort to create certain aromatic flavours for special wines.

A very different innovative or entrepreneurial effort exposed by Gerald Naef from Patina Wines (Orange) is his "Tea Riesling," which he describes as a "happy accident":

> *I think that's the way I approach my winemaking and like to approach it, is thinking the process through. I'm intrigued by different flavours in the wine and just saying, "How would that work with the wine that I'm making? What would it look like?"*

> *Yes, just have fun experimenting with it. The most innovative wine that I've got is the one that I call the 'Sticky Tea Riesling'.*
>
> *That one, I also call it my 'happy accident'. There's a part of its success that I couldn't predict; I couldn't predict that it was going to turn out that way. What I've got is I got tea characters out of the skins of white grapes. Normally from white grapes you do not get tea characters, you get bitter characters from skins, so that part was really unknown for me.*
>
> *What it was, was being a small producer, at the time I made my first Sticky Tea Riesling I only had a basket press. Even some of your bigger wineries, their premium fruit they do in a basket press because it is very gentle, but you never get as much juice from the grapes. In Riesling grapes, the juice is tightly bound inside the flesh and so it's hard to get a lot of juice out of Riesling.*
>
> *With a basket press being a gentle press, I was always only getting probably about two thirds of what I could potentially get out of the grapes. Then I was always dumping the skins out on my wife's compost heap out of the press, knowing that I was losing this 20/30%, whatever, of juice.*
>
> *In 2009, the first one that I made after I dumped them out of the press, I just on a whim decided to just sprinkle some yeast over the top of the skins. Then I shoved them back into the winery for five days, and then I took them out and pressed them again.*
>
> *I really thought that I was just postponing the trip to the compost heap, because I fully expected it to be bitter, but it wasn't; the tannins that extracted from the skins were not bitter at all, they were just drying and then it reminded me of tea.*
>
> *Then a little bit of creativity had to come in, because what do you do with a white wine that tastes like tea? I've never tasted wine like that before (laughter). I thought, "Iced tea is a very good drink."*
>
> *I just decided from that that because I could take some honey characters and some citrus characters in there that I'd just stop the ferment very early while it was quite sweet and make an iced tea wine, basically.*
>
> *It typically is only around 8 or 9% alcohol and the rest... It has a potential, if it fermented all the sugar to alcohol, it had the potential of getting to 12/12.5% alcohol. Yes, so just the combination of sulphur, using a little bit of sulphur and putting it in very cold storage in the freezer room, I'm able to stop the ferment and then just release it as the wine and let in residual sugar.*
>
> *It's a huge success; people love it. (Naef, Orange).*

This is an innovation in the winemaking and targeting a market for lower alcoholic wine. The fermentation process and experimentation resulted in something quite new. The quotation exemplifies what Gherardi and Perrotta [37] call formativeness. Naef is combining playfulness, experimentation, materiality, tactility while forming the object of practice (ibid). In the quote, it is clear than the object of practice is being done as Naef is practising in a sequence where knowing and doing are intertwined and even the name of the wine is subject to its formation.

Gerald Naef is also experimenting with the Riesling, where he experimented with *"stirring lees up to add texture to lengthen the palate."* This was an effort in spite of what they had told him at the university, warning to have Riesling with lees contact, and shows an entrepreneurial and innovative experimentation.

In Domaine la Bégude, we found a surprisingly red rosé and Mr. Tamayan explained that they obtain this through having a longer skin contact than rosés would normally have in the area. *"We have decided not to follow the fashions of the pale rosé."* The very clear strategy to create the skin contact in order to get a different profile from others is an innovation effort.

The experimentation is mostly within the process of stirring (pigeage) and how the skin contact is used for the flavour and for the colour. The most innovative process is the production of the "sticky tea Riesling," as it involved a new process of getting more juice and flavour out of the skins after the first soft pressing. Also it fits well with the effort to cater for a market for low alcohol wine.

It is surprising that the old cement tanks have been re-introduced both for special good wine in Australia and for table wine in South of France. The testing at the lab seems to be supplemented by expert tastings in all areas as the laboratory testing is not seen as sufficient.

5.2 Lighter tasting wines with lower alcohol percentage: managing tactility

Wine is tied to fashion and trends, and to the type of food the wine is paired with that is subject to fashion. This creates a pressure on certain types of wine. This is a clear case of market-driven innovation. In the very hot areas of Australia, the level of alcohol has been pretty high going to 14.5 or 15%. It has been a discussion in both Australia and Spain that the change in food demands wines with less alcohol:

> *Tyrrell's: One of the great things for us in this area is, you know, our Hunter wines are lighter and more delicate; they are cleaner, they are lower alcohol, so they really work with Asian food. Where, say if you go to a McLaren or other areas they are heavier.*
>
> *The thing is, if you eat most Asian food, and the heavy reds are awful there, because they do not, you know, they are not cleaning your palate. They're overwhelming what you are eating.*
>
> *Wine is supposed to complement the food. It's supposed to leave your palate fresh, ready for the next mouthful, not attack it with a baseball bat.... This is what the Americans call "food replacement wines." (Tyrrell's, Hunter Valley).*

This is interesting as a market-driven innovation and relates to the great export adventure of Casella Wines for the American market. They bought a great number of the most popular wines in the US, studied them and worked on making a fit within the price category of $10 ([39], p. 125).

Bodegas Fariña is commercialising many different types of wine under the DO Toro and Vinos de la Tierra. They have introduced wines that "respect the fruit." Rosa Zarza, oenologist bound in Toro, has also used green-harvesting in many of the Toro vineyards where she is working. This is done to develop Toro wines that are less heavy and more fruit-driven to meet a market demand. In this respect, Manuel Perez from Bodegas Fariña is also seriously considering introducing a de-alcoholised (5% alcohol) frizzante style of wine.

This is a new trend, and both in the Spanish and the Australian area, wines traditionally have been quite heavy and with a high level of alcohol. The market has changed as food has changed, and we had several wineries who were concerned with the effort to get high quality and taste but with lower levels of alcohol. This may be difficult and a dilemma, as in hot summers the sugar is rising much more than the flavours.

5.3 Blending

Even if blending is not mentioned in any kind of innovation or differentiation strategy, the blending is mentioned as "the magic moment," which is important for the family owners, for oenologists and for the big decisions on how to make the best wine. It is often stressed that this is an important moment and a practice where the owners have to participate. This is also the place for experiments of how to mix different types of grapes, which is a clear example of tacit knowledge. The magic moment is both relational as it gives owners a chance to raise their opinions and a moment where past experiences are compared with the technical language of the development of the taste of the unfinished wine.

Blending is much more elaborated in Australia. The larger wineries have vineyards in many different places and may use the grapes from hot areas to mix with the cooler climate types of grapes. They always indicate the types of grapes on the bottles, whereas the region is not always mentioned. The blending in France has a lot of restrictions, both regionally and in terms of types of grapes, if they want to be AOC certified:

> *Tyrrell's: We've been through the thing here, the Chardonnay boom, the Sauvignon Blanc boom, which has been unbelievable in this country. I reckon the next one's a white, that's clean and fresh, but got some fragrance and a bit of complexity. We've actually got a wine that's a blend of seven different varieties, one of which is Pinot Noir, decolourised Pinot Noir, which makes it smell like strawberries. It's fantastic. (Tyrrell's, Hunter Valley).*

Blending is also suitable for making sure of a certain level of standardisation. In the case of Casella, the winemakers are specialists, and they use standardisation as a strategy, and it takes a special experience to do this tacit knowledge for sameness or standards:

> *Les Worland: For us, with Yellow Tail, which in Australia is a $10 or $11 product, it has got to taste the same and we do blends of millions of litres, 10, 11, 12 million litres at a time and that has got to taste the same.*
>
> *I mean, I'm a winemaker; I've got a wine making qualification but for these guys, the expertise is to get it to taste the same. Now, for us, we do it with, as I said, 35 other winegrowing areas so we are blending options from cool, warm, hot regions to make that taste the same. (Worland. Riverina).*

In Domaine de Triennes the winemaker Remy Laugier is discussing how his role is in the blending:

> *Laugier: so our barrels where we keep all the cepage (grape variety), grapes are kept separated. Syrah, cabernet sauvignon, especially because we know now our best part, so we keep the different part separated. Then after one year of growing here, we put it*

then back in the tank. And blend after tasting......That is when the chefs comes in and we have to argue, exchange views. (Laugier, Var).

Also in Tempier the family owners and the director are important in the blending, and are always present for the tasting and decisions on how to blend, and what should be the 1-2% other than Mourvèdre.

Rosa Zarza has a strong focus on cellar work; however autonomous in most of the work in the vineyard does always include the owners of a winery in picking the right blend. However, they usually follow her advice, as they trust her judgement and market knowledge.

The blending and the decisions for further treatment are considered the "Magic Moment," where family owners and winemakers are innovation drivers. The effort is very different whether the goal is the best wine, finding the special characteristics of the winery, or as in Casella the purpose is to get the same taste as last year. The "magic moment" is hard to explain, and it is a classic example of tacit knowledge [19, 36], where taste and feelings are a fundamental part of this. The blending has long consequences, as the result is only to be known much later approximately ½-1 year. What is central here is not a proper realisation, as Gherardi and Perrotta [37] found, but rather a respect for the uncertainty of such experimentation that involves living organisms.

6. The use of professional winemakers

In all the Australian cases except one the winemaker is either the owner or employed at the winery. Some of the wineries also take in grapes from small growers and produce their wine for them. Basically, they emphasise the role and skills of the winemaker as the "Chef du cuisine" and the rockstar.

This is less the case in the European samples. When we talk with the daily manager of the winery, who are winemakers in Domaine de Triennes and Domaine la Bégude they also put less emphasis on the winemaker's role. Both Chateau de Roseline and Tempier have oenologists employed, but do not stress their role as in the Australian cases. They more often employ winemakers as consultants, and their names are not mentioned on the bottle or homepage.

In Domaine Jas d'Esclan the winemaker is a consultant who has experience from other wineries and is not involved in the vineyard. They stress that the knowledge of the markets is used in the winemaking. Val d'Iris started using the consultant a lot but use him now on a less frequent basis.

Also in Spain the oenologists are often working outside the firm. The purpose of hiring an external oenologist is twofold. To make the blend and give guidance and very importantly to provide knowledge about the market to the viticulturist:

Interviewer: And the oenologist he, since how long is he working with you? As a starter you only had the knowledge of your father, right?

Arrévalo: Well, we have always had people who helped us, from other wineries or oenologists, but little by little, after 20 years of making wine you get the ideas, your way of making your wine and well, what I missed in terms of chemistry, the thing about laboratory analysis well that is supplemented by a person. And a bit of market knowledge, how the tendencies are, or what I want, that is a bit what I do. (Arrévalo, Rueda).

The winemaker is the expert on the whole process of growing vine, grapes, fermentation, blending and the chemical processes during the whole production process. In Australia the winemaker is the Chef and the hero, and in some cases nearly a "Rockstar." The role is highlighted and the name may be on the homepage to show his/her reputation. In France and Spain, however, she/he is an expert consultant. Australia has been developing wine educations, and wineries have this competence in house in most cases. The more traditional wineries in Europe stress the power and competence in house, but the need for some consultant expertise, who also bring some experience from other wineries and the market. In two cases in France, the winemaker is employed and running the winery. Generally, this is a role closely tied to the winery, even if it is an external consultant, she will be part of the local practice. But even with professional assistance, the tacit knowledge of both owners and oenologists is emphasised. To conclude, the external winemaker at times goes beyond her role as a technical advisor to become part of the learning organisation of the winery.

7. Discussion of innovation and entrepreneurial practice

Why is the wine field so special as compared to other agricultural areas? The wineries seem to have a very strong feeling of identity with the quality of their wine and often with the region as a hint of terroir or a regional-based quality stamp like AOC/DO. There is very limited reference to customers and marketing. The way of identifying what is quality is not uniform, and in their stories, the agricultural cultivation of grapes plays a very important role in the wine production. The selection of grapes and the treatment of grapes and use of water and fertilisers are emotionally explained; why they are so important. Only when asked, the process of winemaking in the winery is described.

We have tried to illustrate where in the wine production innovations or entrepreneurial practice is taking place. We have shown several examples of entrepreneurial practice and experiments to illustrate the research question, "how does constant experimentation contribute to the practice of innovation in wineries?"

Most of the existing innovation analysis is based on the regional support structures. The R&D and oenological development support in Australia seemed to be important, and the development of institutional competence and regional oenological expertise appear to have an impact and are more dominant than referred in our European cases. In a later article, Aylward et al. [40] emphasise how South Australia "has successfully integrated the core ingredients of viticulture, oenology and the organisational and marketing requirements into a highly evolved mix of innovation activity" ([40], p. 46), and that Victoria and New South Wales have less developed clusters (ibid.). This is also the main focus for Farias and Tatsch [1] in Brazil, where they find that the Serra Gaucha region has a well-developed innovation structure, yet when asked, respondents claim that 95% of the innovations are conceived within the wineries.

We have found that drivers for innovation might be at the local level (as also indicated by Farias and Tatsch [1]), such as the scientific laboratory of Matarromera studying how to treat the soil, how to use the skin of grapes, to experimentation with skin contact and how the skin is involved in the fermentation.

In relation to innovation and need for new methods, there are many informants who emphasise the uncertainty of wine production, both in relation to the climate and seasonal changes, and the winemaking. For understanding the processes of innovation and experimentation, this time perspective is fundamental. You may be able to adjust something, and you may blend or add other flavours, but experimentation in the early stages is quite determining for the later result. Here the entrepreneurial actions are revealed in the practice of the wine producers. With the high uncertainty of the outcome of experiments, it resembles more drastic innovation processes, even if it is based on existing knowledge and competence.

It is a challenge to ask for innovation and change in a sector so much dominated by the great stories of tradition and quality, and where very many of the owners are in old family firms, where the great grandfather and the grandfathers have been establishing the winery. Family and tradition have been quite dominating in the French and Spanish firms, but also in some of the Australian firms, where some present owners are fourth generation. This emphasis on the traditional values and qualities however does not mean that they are not developing and experimenting. They act like entrepreneurs and try new methods. Some are very well organised and work on technological changes, and some are playful in their experimentation, trying new ways and new types of wine in a process that resembles entrepreneurship by improvisation [22].

Even if they do not want to relate to the concept of innovation, the effort to "distinguish and differentiate" themselves from other local winemakers is easily understood, and gives a good response on what they do, as they compare their wines to others'.

The wineries build on existing knowledge in the tradition and a strong identity as winemaker, but also lean on chemical knowledge as, e.g., the oenologist as the knower. Most innovation processes are clearly incremental and build on existing knowledge but with some new methods or adaptation of technology and product, which is seen as a personal strategy and effort of experimentation. But a few are experimenting, such as Naef when he creates lees contact with Riesling, against the university knowledge.

When we stress the entrepreneurship and the practice in wine production, it is because we have been asking about "how they do differently," i.e., their practice of experimenting. It is not innovation at a macro or regional level but an effort to make better wine in all the parts of the complex process from agriculture to winemaking. It is the individual level strategy and practice we wanted to address. In this perspective, we have found experimentation and playfulness in an entrepreneurial approach, which has been convincing.

Peter Drucker stresses that *"Entrepreneurship is neither science nor arts, it is practice"* [38]. Johannisson concludes "hands-on action and concrete practice for very good reasons dominate everyday life in entrepreneurial firms" [41]. If entrepreneurship is practice and the effort to be open to serendipity in an iterative process, such as also in the process perspective in Sarasvathy's theoretical approach on entrepreneurship as effectuation and practice [27], then it fits better with the behaviour and explanations we observe in the field. It is a step-by-step experimentation as a strategy for innovation and improvement, which answers our research question. In this way, the present study adds to the knowledge in the field on how playfulness and experimentation form the nucleo of innovation in wineries.

7.1 Experiments and playfulness under conditions of uncertainty

Wine production is very complex and involves many processes from agriculture, to fermentation to blending and ageing. It is a process of risky experiments, as the time period to get results is so delayed.

The high level of uncertainty, which is embedded in innovation and in climate dependent production, is at play here. Many wineries do not have irrigation and are thus highly dependent on the weather. Also the efforts in the fermentation, as Naef puts it: "*Wine is a complex soup of molecules.*" Thus, they can experiment with certain wines from time to time but have to keep most of the wines stable to secure a permanent income. But we find that they all talk about the experiments with such engagement, as they find it highly challenging and exciting. It appears to be a playful experimentation, and the happiness if it works is obvious. Such as describing the "happy accident," this is not an accident but an effort to try to get a new taste out of the skins after the first pressure. This is a clear innovative experimentation, which created a new successful product.

The winemakers in Australia compare themselves to "Chefs" in the kitchen, but they emphasise: "*winemakers, we're kind of like chefs with a lot of patience; instead of putting it in the oven and the next day or the next hour learning how it turned out, we've got to wait for a whole year*" (Naef, Orange). This is underlining the emphasis on uncertainty and the perception of entrepreneurial practice.

Ann D'or experimented with longer fermentation with the skin in the red wines, and managed to get a much stronger and concentrated red wine, than the other local wineries. Her special experiment takes more effort in the winemaking process to get her high quality-high price wine: "Le Grand Foudre," and she only does this every second year. The way the technology and methods are discussed is very much like technological actors are understood in an Actor-Network-Theory perspective [42].

The many laboratory tests on the grapes and the juice in the tanks are in all cases supplemented by tastings and visual judgement. When winemakers explain how they experiment or innovate a good procedure, it is attached to the judgement of the type "I can do it again" in the sense of Schutz [43]. They need new methods to evaluate sugar content and acidity, but they all talk about the necessity for tasting and their use of tacit knowledge and tactility. In the blending, it is a sacred moment to estimate the quality, sense the feeling and colour and find the "magical moment." The family and owners are always part of this. It is not left to the oenologists, whether employed or consultants.

Experiments in vines and types of grapes were found in all areas. This is a long process as growing new vines takes time to mature for proper grape production. But quite many do experiment with new types of grapes; first, within a small area, and if it works, it can be expanded. The efforts are analysed well before comparing the local soil and climate to find other types of grapes that will fit the area. This can be the Gewürztraminer (Sweetapple), the Riesling (Naef), the Viognier (Dolle) or the Mourvèdre, which is known to be best close to the coast (Christine). These types of grapes are not found before in the area, and it is a deliberate strategy to create a different profile of wine than the local competitors. However, most of the cases are experiments, and in the description, they tend to a certain playfulness in the exploration of new types of grapes for blending.

All in this sample of the smaller wineries, which are in the middle range of wine, i.e., good wine, but not the cheap table wine or the extraordinarily very expensive wine, emphasise the effort to reduce chemicals and pesticides. Many are very close to organic production but hesitate to call it such, as they perceive the US market as "impossible to live up to the standards." In some cases, the owners say that it does not contribute to the marketing to be organic. We find many efforts of experimentation and improvisation within this field, but also a lot of frustrations as they find too many barriers and limitations and that what they can do is not sufficient for their perception of what should be organic production. The whole discussion of organic is more like a very competent assessment of the many constraints in organic production and the way they try to create a practice that matches their philosophy of organic production. However, they are not very happy with the existing solutions, and they accept the necessity still to use sulphur, even if they are experimenting to find new solutions.

Some, however, work on tradition, where they never went into the use of chemicals, such as Jas D'Esclan. Traditional agriculture from before the Second World War was organic. In this way, tradition and novelty in demand are coupled.

The improvisation and playfulness in most of the aspects of winemaking are closely tied to routines and tradition in most parts of the winemaking stories. There are many routines and similar traits of winemaking, but how they include experiments and innovation often in a "corner" of their practices is usually told with a high level of engagement and humour, as if these aspects are the interesting part of the winemaking process. The playfulness of experimentation and improvisation forms the creativity of the entrepreneurs in winemaking. Just like in the study of Gherardi and Perrotta [37], we have found empirical evidence for playfulness in how the winemakers experiment in every phase of viticulture and winemaking.

We find that most innovation efforts in the data are examples of playfulness in a craftsmanship setting. The knowledge that is needed for being playful in such a setting has strong connotations of tacitness. The tension between explicit and tacit knowledge in the innovation process is seen in the strive towards educational efforts and me-too innovation and yet most of the innovation process is described as being playful and experimenting in a craftsmanship setting where only actors that are immersed in the community of practice are involved [19]. The question whether to have a winemaker employed at the winery or to have an external consultant is rephrased into: Do we want an oenologist working closely with the winery in both the field and the cellar? The wineries that stress playfulness and experimentation are all nodding affirmatively to this question. Only by being a member of the community of practice can the winemaker gain the tacit knowledge that is essential for improvisation and playfulness. Because the knowledge is so tacit, the "magic moment" needs people with experience of repetitive practice to form patterns to see if the blends are typical or deviant [36].

The engagement and efforts to do something different, even without knowing the result, are clear in many of the quite entrepreneurial wineries. They explain that the result is extremely uncertain but try to make new combinations or methods. When we call it playfulness, it is because of the eager engagement of the entrepreneurs and their identity feeling as entrepreneurs [44], as well as the way they describe the process as exciting or playful. The way they proudly describe the experiments and their joy when they succeed makes us use the concept of playfulness. In this sense, the entrepreneurs in the wineries expose traits of *homo ludens*; while, for a while, hiding their *homo economicus* [11].

7.2 Technology and process innovation

The range of technological development and innovation is very large. One firm in Spain participates in very large EU projects both in agriculture and in developing other products from the vine. One firm is very innovative in the water and recycling of chemicals. But most are to be considered early adopters and have different ways of organising and creating smaller changes. New technology such as steel containers to control temperature is found everywhere now, but old cement open containers may supplement, even in very advanced wineries, such as Triennes in Var, France or in Westend Estates Wines (Riverina), where the 60-year-old open cement containers were used for premium wine. The open cement containers allow for the possibility to use tactility and stir the juice with the skins and watch the process. They emphasised that it is not rocket science, yet often unpredictable.

Some of the owners and employees are trained in technology and focus on technology and machines. This is true in some of the cases in Australia and Spain, where viticulturists are developing machinery. Also, the effort to innovate the water cleaning system is found in one of the very large producers (DeBortoli in Riverina). This is a case of a clearly defined problem, and a planned large-scale radical innovation project, which is recognised by Government funding. It covers a range of method development in cleaning water and recycling potassium and represents a very different and well-planned method to innovation. Such process innovation within climate, water and recycling chemical additives is quite exceptional. It is only possible in very large wineries, like in bodegas Matarromera, Ribera del Duero and DeBortoli, Riverina, where an engineer specialising on this has the possibility of running such a large innovation project.

Perez has turned around some of the challenges in the harvesting, where he identifies a problem of time from harvesting to the tanks in the winery. He has changed the organisation of the harvesting and transport process by moving the de-stemmers and tanks to the field to shorten the time from harvesting to tanks. He has developed the machinery adaptation himself (Perez, Toro). This is a very original and innovative organisational process, involving some technological adaptation and renewal as well. It is a good example of process innovation, where Bodegas Fariña has identified a problem of grapes crushing and starting the fermentation while still on their way to the winery after being harvested. Technology-minded entrepreneurs use their engineering background for measuring water humidity along the roots of vines in different layers of soil (Perez and Zarza), and as we can see, the new technology may reveal for the oenologist whether the grape-grower is irrigating or not. Bodegas Matarromera had big European research projects on how to create irrigation by drops in the most efficient way.

These types of technological innovations are important and technology-driven. They are very different from some of the other locally developed experiments.

The irrigation issue is seen to vary in different regions and by different growers. In some of the areas in Orange, in France and Spain they emphasise that they do not irrigate at all or in certain periods. For some, this is because they do not have water access, and for some, it is an ideology or, in several cases, such restrictions by the AOC/DO. In these cases, the distance control of humidity becomes an important innovative method.

Both activity theory and sociology of translation (or Actor Network Theory) recognise the importance of describing the network features of action, but they disagree in how they come about. Activity theory focuses on culture and historical paths, whereas the sociology of translation gives importance to humans and non-human elements ([20], p. 682). In the field of wine, it is imperative to give explanatory power to both forms of causality. Machinery and technology in the vineyard and the cellar shape the way knowing in practice is constituted in the community of practice. Simultaneously, the tradition of how to do things strongly influences how knowing in practice is done. This is more dominant in the wine sector than other agricultural sectors because of the strong positive associations a winemaker or viticulturist can make by connecting to local tradition. A study design of practice in the wine sector would have to cater to both research traditions.

8. Conclusion

The innovation and entrepreneurial practice in the wine sector is a very complex pattern, everywhere, with aspects of experimentation and entrepreneurial practices in different parts of the wine production. As we do not have a representative sample in the three countries, standardised comparisons are meaningless.

There has been an emphasis among the wineries to focus on the practice in the viticultural aspects, and they describe how they deal with different types of experiments with different kinds of grapes, which are unusual in the area, or working on organic or nearly organic production. The interviews have emphasised how the interviewees focus on the doing of practices rather than the intentionality of those practices [20]. The effort to do organic production is full of challenges, and even with these challenges of the technology, very few use organic features in marketing.

Our results lean towards innovation that could be labelled as "creative" yet tactile, where existing practices are put together in new relational patterns [11]. The experimentation is quite entrepreneurial and an interesting feature in such a traditional sector. The practice of wineries is in focus, and the entrepreneurial practice seems to be a promising way of explaining innovation in the wine sector. Such practice-based explanations could include empirical data with longer field observations and focus on the relation between knowledge, objects and participants in wine-growing and -making.

In our material, winemaking is represented by the interviewees in such a way that the knowing and doing of wine practices are inseparable [20]. When the viticulturist speaks about managing sunlight to get the right amount of leaf shadow to the grapes, it is a practice that is embodied in the work, where knowledge is embedded in the culture and *savoir faire* of doing and undoing the same practices over and over again with different grapes under varying climatic circumstances. This is learning by doing in an improvisation.

Our contribution is to show how experimentation and practices of innovation take place in the wineries. This is an illustration of how the practice of innovation is performed at the individual level in the innovation process.

Wine production is a very complex process; experimentation and innovations can be found in several aspects of the production process, from viticulture to winemaking. Experimentation is to be found in the cellar, where the winemaker engages in playfulness to improve the wine. We have seen that the quality wineries manage their portfolios of wine in such a way that some wines are cash cows, a certain part of the production is saved for playing around with. There is never a stand-alone rational management in the wineries. Playfulness and improvisation are always an integrated part of day-to-day practices when the owners and winemakers are engaged in the long quest for a wine commercial.

There is a need for further research on strategies to act entrepreneurial in wine production and the regional differences between these strategies. We still know too little about practice, improvisation and playfulness in wine production.

Author details

Christian Poulsen[1]* and Mette Mønsted[2]

1 Department of Management, Society and Culture, Copenhagen Business School, Copenhagen, Denmark

2 Department of Management, Politics and Philosophy, Copenhagen Business School, Copenhagen, Denmark

*Address all correspondence to: chpo@kea.dk

References

[1] Farias CVS, Tatsch AL. The Brazilian wine industry: A case study on geographical proximity and innovation dynamics. Revista de Economia e Sociologia Rural. 2014;**52**(3):515-532

[2] Castellano S, Khelladi I. How French wine producers use open innovation to gain and manage their legitimacy. Journal of the Knowledge Economy. 2015;7:155-171

[3] Doloreux D, Lord-Tarte E. The organisation of innovation in the wine industry: Open innovation, external sources of knowledge and proximity. European Journal of Innovation Management. 2013;**16**(2):171-189

[4] Dressler M. Innovation management of German wineries: From activity to capacity—An explorative multi-case survey. Wine Economics and Policy. 2013;**2**:19-26

[5] Edquist C. Systems of innovation: Perspectives and challenges. In: Fagerberg J, editor. The Oxford Handbook of Innovation. Oxford Academic; 2005. pp. 181-208

[6] Lundvall BÅ. National Systems of Innovation. Towards a Theory of Innovation and Interactive Leaning. London, Wellington House: Anthem Press; 1992

[7] Aylward DK. A documentary of innovation support among New World wine industries. Journal of Wine Research. 2003;**14**(1):31-43

[8] Aylward DK, Turpin T. New wine in old bottles: A case study of innovation territories in "the New World" wine production. International Journal of Innovation Management. 2003;7(4):501-525

[9] Rauch A, Frese M, Wang ZM, Unger J, Lozada M, Kupcha V, et al. National culture and cultural orientations of owners affecting the innovation-growth relationship in five countries. Entrepreneurship and Regional Development. 2013;**15**(9-10):733-755

[10] Svare H, Gausdal AH. Strengthening regional innovation through network-based innovation brokering. Entrepreneurship and Regional Development. 2015;**27**(9-10):619-643

[11] Johannisson B. Towards a practice theory of entrepreneuring. Small Business Economics. 2011;**36**:135-150

[12] Taura ND, Watkins D. Counteracting innovative constraints: Insights from four case studies of African knowledge-intensive metalworking and automotive clusters – 'The Akimacs'. Entrepreneurship and Regional Development. 2014;**26**(3-4):313-336

[13] Nonaka I, von Krogh G, Voelpel S. Organizational knowledge creation theory: Evolutionary paths and future advances. Organization Studies. 2006;**27**:1179-1208

[14] Lukacs P. Inventing Wine: A New History of One of the world's most Ancient Pleasures. New York: W.W. Norton & Company; 2012

[15] McIntyre J. Inventing wine: A new history of one of the world's most ancient pleasures. Journal of Wine Research. 2015;**26**(3):243-249

[16] Bessant J, Tidd J. Innovation and Entrepreneurship. 3rd ed. Chichester: Wiley; 2015

[17] Lenzi C. Smart upgrading innovation strategies in a traditional industry:

Evidence from the wine production in the province of Arezzo. Regional Science Policy & Practice. 2013;**5**(4):435-452

[18] Gartner WB, Stam E, Thompson N, Verduyn K. Entrepreneurship as practice: Grounding contemporary practice theory into entrepreneurship studies. Entrepreneurship and Regional Development. 2016;**28**(9-10):813-816

[19] Brown JS, Duguid P. Knowledge and organization: A social-practice perspective. Organization Science. 2001;**12**(2):198-213

[20] Gherardi S. To start practice theorizing anew: The contribution of the concepts of agencement and formativeness. Organization. 2016;**23**(5):680-698

[21] Cook SDN, Brown JS. Bridging epistemologies: The generative dance between organizational knowledge and organizational knowing. Organization Science. 1999;**10**(4):381-400

[22] Cunha MP. Entrepreneurship as decision making: Rational, intuitive and improvisational approaches. Journal of Enterprising Culture. 2007;**15**(1):1-20

[23] Cunha MP, Viera da Cunha J, Kamoche K. Organizational improvisation: What, when, how and why. International Journal of Management Review. 1999;**1**(3):299-341

[24] Weick KE. Improvisation as a mindset for organizational analysis. Organization Science. 1998;**9**(5):543-555

[25] Vera D, Crossan M. Improvisation and innovative performance in teams. Organization Science. 2005;**16**(3):203-224

[26] Kuhn T, Jackson MH. Accomplishing knowledge: A framework for investigating knowledge in organizations. Management Communication Quarterly. 2008;**21**(4):454-485

[27] Sarasvathy S. Effectuation: Elements of Entrepreneurial Expertise. Edward Elgar Publication; 2009

[28] Heckathorn DD. Respondent-driven sampling: A new approach to the study of hidden populations. Social Problems. 1997;**44**(2):174

[29] Amsteus MN. The validity of divergent grounded theory method. International Journal of Qualitative Methods. 2014;**13**(1):71-87

[30] Corbin J, Strauss A. Grounded theory research: Procedures, canons, and evaluative criteria. Qualitative Sociology. 1990;**13**(1):3-21

[31] Glaser BG. Basics of Grounded Theory Analysis.: Emergence Vs Forcing. Mill Valley, CA: Sociology Press; 1992

[32] Beverland M. Branding authenticity in the luxury wine trade. Journal of Business Research. 2006;**59**:251-258

[33] Potter J, Hepburn A. Qualitative interviews in psychology: Problems and possibilities. Qualitative Research in Psychology. 2005;**2**(4):281-307

[34] Tryggestad K, Georg S. How objects shape logics in construction. Culture and Organization. 2011;**17**(3):181-197

[35] Poulsen C, Mønsted M. Identity markers for wine producers: Terroir and beyond. Journal of Wine Research. 2020;**31**(3):194-217

[36] Tsoukas H. How should we understand tacit knowledge? A phenomenological view. In: Easterby-Smith M, Lyles MA, editors. Handbook of Organizational Learning and Knowledge Management. 2nd ed. Chichester: Wiley; 2011. pp. 453-476

[37] Gherardi S, Perrotta M. Between the hand and the head: How things get done, and how in doing the ways of doing are discovered. Qualitative Research in Organizations and Management: An International Journal. 2014;**9**(2):134-150

[38] Drucker PF. Innovation and Entrepreneurship. Practice and Principles. London: Pan Books; 1985

[39] Dufour Y, Steane P. Building a good, solid family wine business: Casella Wines. International Journal of Wine Business Research. 2010;**222**:122-132

[40] Aylward D, Glynn J, Gibson B. SME innovation within the Australian wine industry: A cluster analysis. Small Enterprise Research. 2006;**14**(1):42-54

[41] Johannisson B. Disclosing everyday practices constituting social entrepreneuring – A case of necessity effectuation. Entrepreneurship and Regional Development. 2018;**30**(3-4):390-406

[42] Latour B. Reassembling the Social. An Introduction to Actor-Network-Theory. Oxford: Oxford University Press; 2005

[43] Schutz A. The Phenomenology of the Social World. USA: North Western University Press; 1967

[44] Leitch CM, Harrison RT. Identity, identity formation and identity work in entrepreneurship: Conceptual developments and empirical applications. Journal of Entrepreneurship and Regional Development. 2016;**28**(3-4):177-190